Section 1

Physical Quantities, Units & Measurement

1.1 Scalars & Vectors

A scalar is a quantity described fully by a size, or magnitude, alone; i.e. a numerical value. A vector is a quantity possessing both a magnitude and a direction. Arrows are conventionally used to indicate vectors and point in the direction in which they act. Often, if the vector acts on an object in an inclined plane, the angle is given so the vector is fully defined:

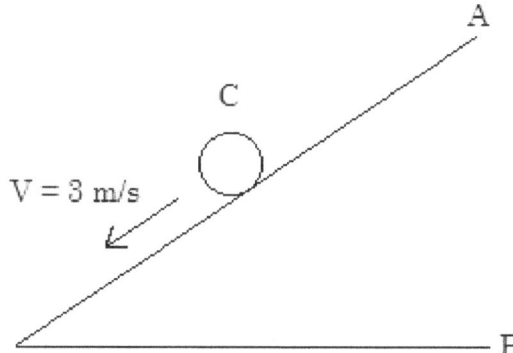

Figure 1.1

The velocity vector v indicates that C moves with a speed of 3 m/s in the direction shown.

Certain scalars and vectors are often, but mistakenly thought of as being interchangeable. In reality, the vector elements of this set all have a scalar counterpart or analogue, which may be identical in size to the vector, but is obviously not the same since it lacks direction. Also note that a vector multiplied by a scalar produces a vector. The table below clarifies this with examples.

Vector Quantity	Units	Scalar Analogue(s)	Units
Displacement	m	length, distance	m
Velocity	m/s	speed	m/s
Weight	N	mass	kg
—	—	time	s
Acceleration	m/s^2	magnitude of acceleration	m/s^2

Table 1.1

Notice that time is always a scalar, and acceleration is always a vector. See also that displacement and velocity have the same units as their scalar analogues, yet weight has different units from mass. This is because weight, or more generally force, is equal to mass x acceleration, according to Newton's Second Law. In the case of weight, this acceleration is caused by gravity, the mass of the earth bending space-time. Here, acceleration is represented with the letter g, standing for the gravitational field strength, which is equivalent to an acceleration of 9.81 m/s^2. Since mass times acceleration is a scalar times a vector, the result is a vector, although the units of the vector produced now differs from the scalar analogue. Accordingly, 1 Newton (N) is equivalent to 1 kilogram (kg) times an acceleration due to gravity, termed g, of 9.81 m/s^2.

It is important to distinguish between weight and force. Weight is just a particular type of force, measured in the direction 'straight-down' to the centre of the earth. Forces as a result of masses accelerating can occur in all different directions, but conventionally, the force acting straight down on all objects is termed their weight. The weight of an object can vary depending on the acceleration it feels, for example an object on the moon will be 'less heavy' than it would be on earth, although the mass of the object remains the same.

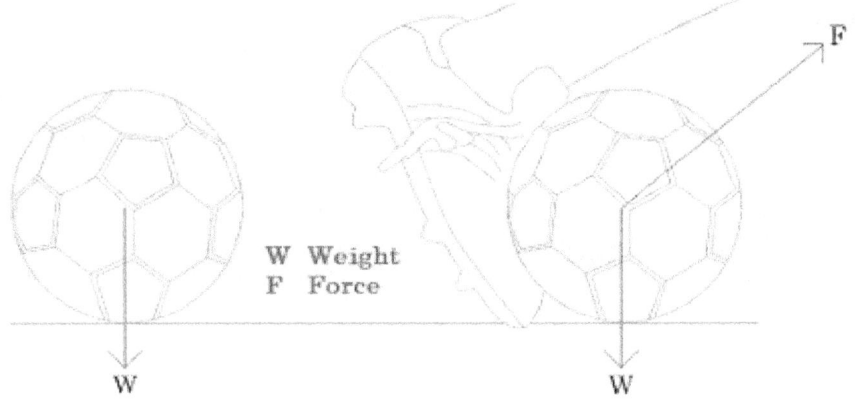

Figure 1.2

It is important to know the distinction between weight and force.

If a ball is kicked from the ground in still air, the direction in which it will travel can be determined by resolving the forces acting on it, W and F, to produce a resultant force. If the weight and force vectors W and F are drawn to scale then the resultant can be determined. Forces which act in different directions are called inclined forces. To find the resultant a parallelogram of forces is drawn. A line originating from the arrow point of W is drawn parallel and equal in length to F, and a line originating from the arrow point of F is drawn parallel and equal in length to W as follows:

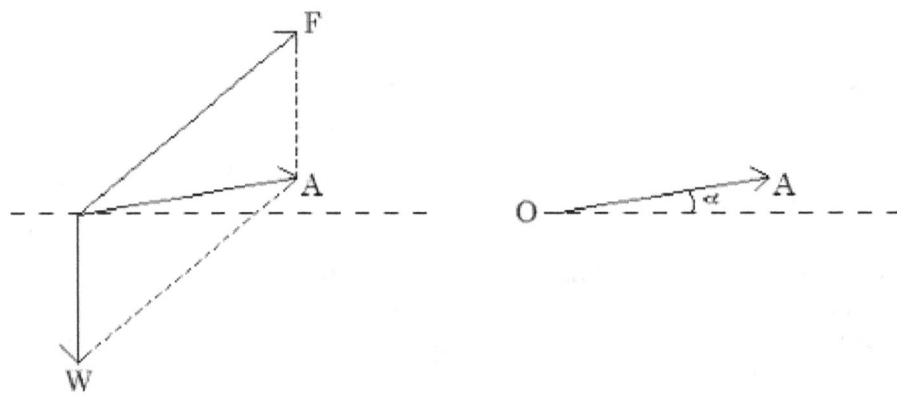

Figure 1.3

A parallelogram of vector forces, left, and the resultant, right.

The point at which the two dashed lines meet is the end point of the resultant, and a line drawn from the origin O to this point forms the resultant vector OA. If F and W are drawn to scale at a scale of 1cm to 30N, then the resultant force OA can be determined by measuring its length in cm and converting to Newtons by multiplying by 30. The angle α at which OA is inclined to the horizontal can be found by measuring with a protractor (note that this diagram is NOT drawn to scale).

The examples below, all drawn to scale, show the resultant force OA, of OB and OC. The scale used is the same for each, 1 cm = 20 N. Determine the resultant forces OA, and the angles α and γ in each case.

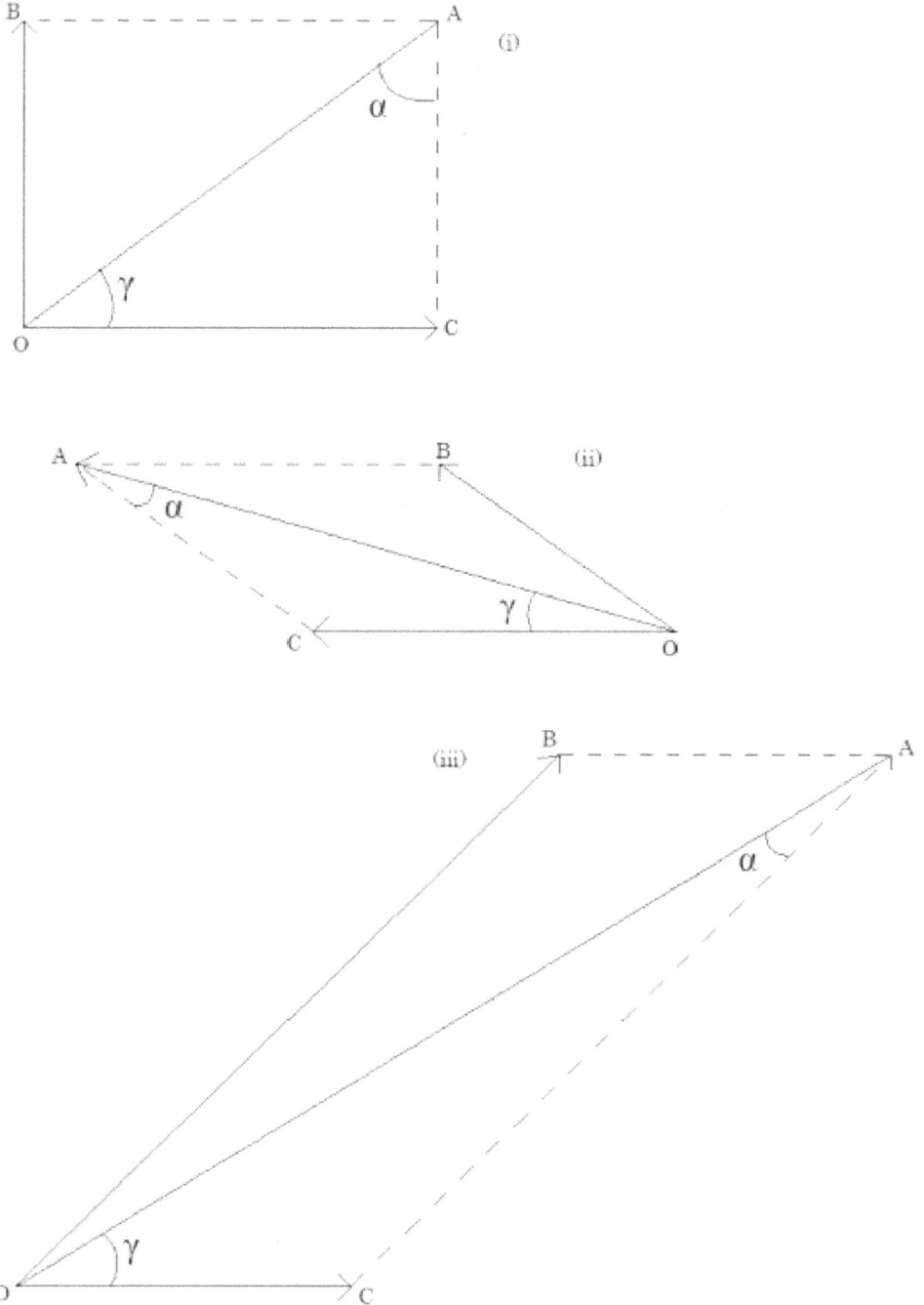

Figure 1.4

A series of parallelograms of forces where force OA, α and γ are to be found in each case.

(Answers: (i) OA = 138 N, α = 53°, γ = 37°;   (ii) OA = 164 N, α = 20°, γ = 16°; (iii) OA = 276 N, α = 13°, γ = 32°)

1.2   Measurement Techniques

Length

A number of instruments exist for measuring length. The most common is the measuring tape, or metre rule. A measuring tape is suitable for measuring a wide variety of lengths, ranging from a few centimetres to several metres. For most everyday prurposes, and many laboratory experiments, a rule or tape is sufficiently accurate.

Provided that a rule or tape has been correctly calibrated against an approved standard, then the best absolute measurement possible with it is one half of the smallest increment (division) marked on the scale. For most rules or tapes the smallest increment is 1 mm, so the most accurate measurement achievable is to plus or minus 0.5 mm. This figure is known as the resolution uncertainty. A figure can only be given with this much accuracy if the quantity being measured is stable or fluctuates sufficiently slowly with time to permit such a measurement. Otherwise the quantity measured must be given to an accuracy commensurate with the reading uncertainty, which will be more uncertain than that of the resolution.

Calipers are another simple device used for measuring lengths. There are two types, inside calipers and outside calipers. Both are extremely useful, since they can accurately measure the inner and outer dimensions of complex objects. The calipers are set so that their edges touch the inner or outer surface of the part to be measured. This length can then be measured off against a ruler to an accuracy of one half of the smallest graduation, typically 0.5 mm.

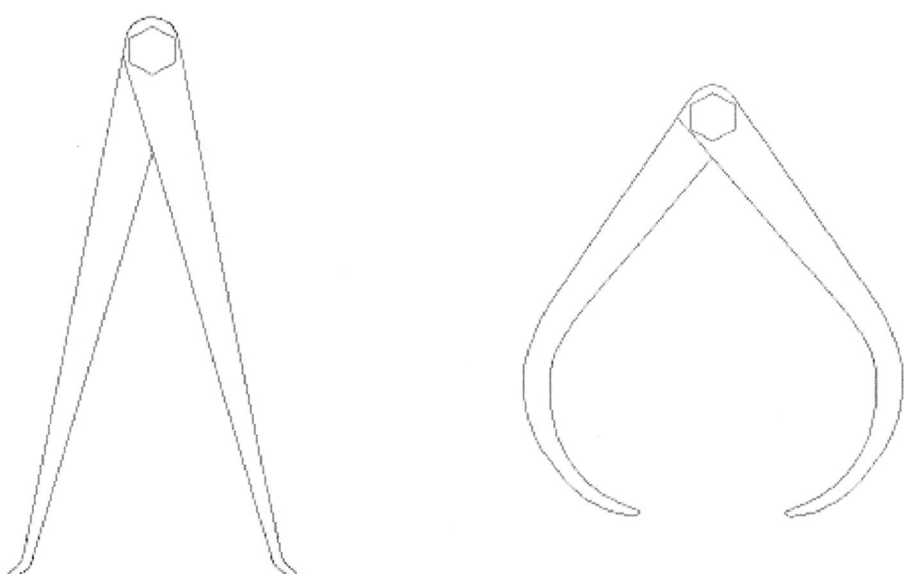

Figure 1.5

Inside calipers, left, and outside calipers, right.

A Vernier calipers is a very accurate measuring instrument whose resolution uncertainty is 0.05 mm. It can combine both inside and outside calipers into a single unit. The movable jaw has a scale on which graduations are placed 0.9 mm apart giving 10 intervals. The 0 and 10 on this scale align exactly with the 0 and 9 mm graduations on the main scale when the jaws are closed.

When the jaws are square to the object being measured, the sliding scale is examined to see where the zero mark falls. Wherever it falls, it is always rounded down to the nearest graduation, in millimetres, on the main scale. Whichever mark on the sliding scale coincides most closely with a mark on the main scale gives the second part of the reading in tenths of a millimetre.

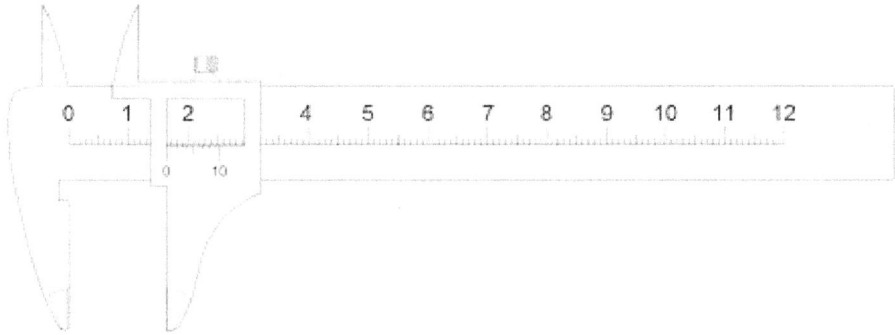

Figure 1.6

A typical set of Vernier calipers.

A micrometer is a device used for measuring small and narrow objects, such as the diameters of wires and rods. The object is placed in the horseshoe between the spindle and its end point, and the barrel turned until the object is touching both the spindle and end point, on either side. The end of the stem is a nut threaded so that it matches the 0.5 mm pitch of the screw, which forms the centre of the barrel. A ratchet is often used to prevent overtightening of the spindle, by preventing further forward motion once the spindle has contacted the object being measured. There are 50 graduations of 0.01 mm each on the barrel, and a linear scale on the stem, with graduations 0.5 mm apart, such that one rotation of the barrel corresponds with a linear movement of one interval on the stem. The point at which the two scales intersect gives the measurement reading.

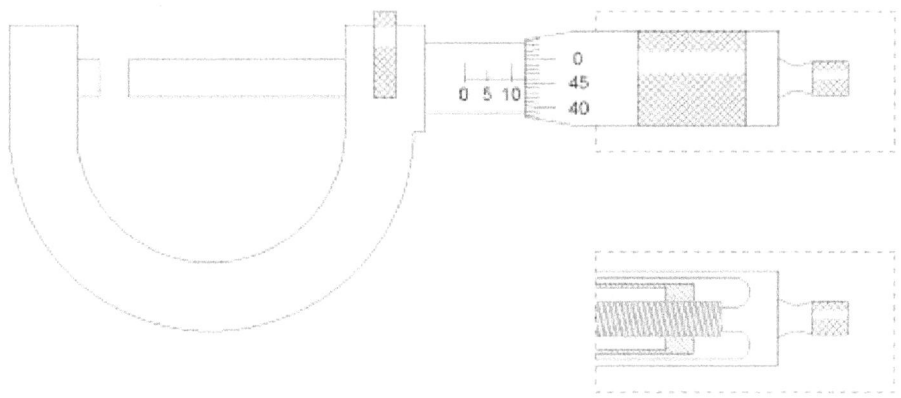

Figure 1.7

A typical micrometer.

Note that for instruments such as Vernier calipers and micrometers, a zero error may be present, whereby the scale does not read zero when the jaws or spindles are closed. If this is the case, the device may still be used, but it should be remembered to either add or subtract the initial difference, based on whether the instrument records zero as a negative or positive value respectively at what should be the zero position.

Time

Digital stopwatches and clocks are often used to measure time intervals in laboratory experiments. The read-outs of digital stopwatches are accurate to 1/100 th of a second and clocks are accurate to within a second. Human reaction time is on average about 0.2 seconds however. This is the limiting factor in the case of timing measurements made with digital stopwatches, whose readings should be considered accurate to +/- 0.1 s at best.

For clocks, the smallest increment of time measurable is one second, so the accuracy of a measurement is limited to +/- 0.5 s.

When timing short and repeatable events in the laboratory, several recordings should be made and averages taken. Any timing(s) markedly different from the others recorded should be attributed to miscounting and therefore repeated.

When measuring oscillations such as the swing of a pendulum, a fiducial mark should be used. This is a reference point (notch or mark), and an indicator to make a count as the object swings by on each cycle.

Section 2

Kinematics

2.1   Speed, velocity & acceleration

Speed is a measure of the rate at which something is travelling. It is the gradient of a distance-time graph. The average speed s, in the time taken to travel a distance d, is given by:

$$s = \frac{d}{t} \qquad (2.1)$$

$$\text{i.e. average speed} = \frac{\text{distance}}{\text{time}}$$

Velocity is the change in displacement divided by the time taken. Acceleration is the change in velocity divided by the time taken:

$$a = \frac{v - u}{t} \qquad (2.2)$$

$$\text{i.e. acceleration} = \frac{\text{final velocity - initial velocity}}{\text{time}}$$

2.2   Graphical Analysis of Motion

Speed may be calculated from a distance-time graph at any point by finding the gradient there. First, a line is drawn parallel and coincident to the curve at the desired point. This is the tangent. A base and perpendicular are then drawn at 90° to each other, forming a right angled 'gradient triangle'. The tangent is the hypotenuse. The length of the perpendicular is divided by the length of the base to give the gradient and a value for the acceleration. If the curve is increasing from left to right, the gradient is positive. Otherwise it is negative.

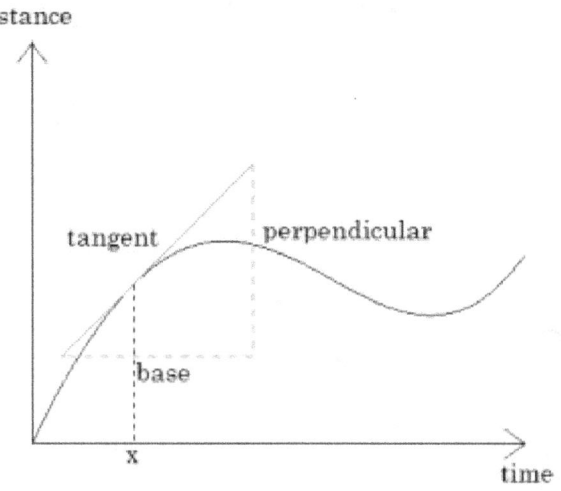

Figure 2.1

A distance-time graph showing a gradient triangle constructed to find the gradient at x.

$$\text{speed at time x} = \frac{\text{length of perpendicular}}{\text{length of base}}$$

A speed-time graph is plotted in the same fashion as a distance-time graph. The gradient of a speed-time graph gives the size of the acceleration. The plot below illustrates the speed-time history of a body as it undergoes different motions. Total distance travelled is equal to the area under the graph.

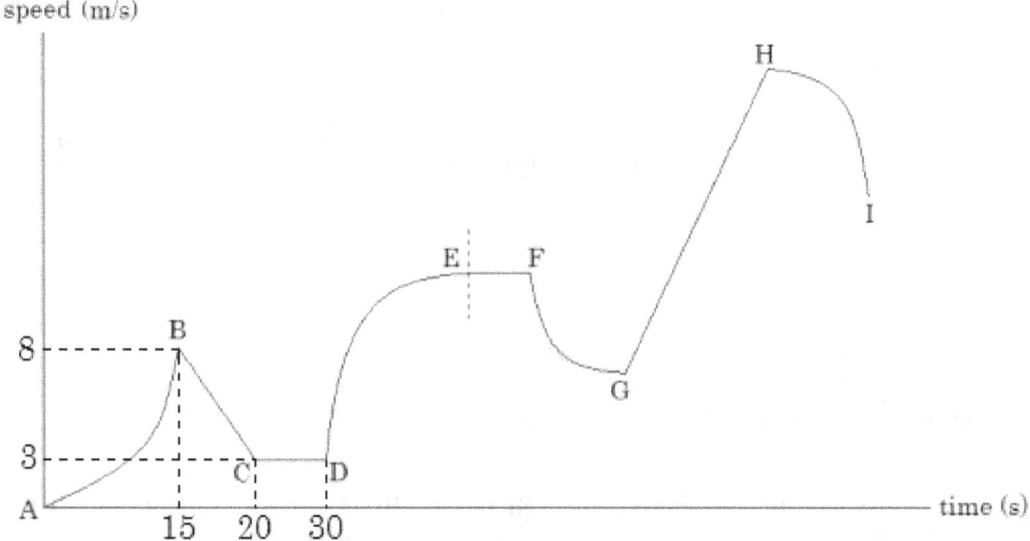

Figure 2.2

A speed-time graph illustrating the different types of motion between the lettered regions.

The various regions represent as follows:

AB	increasing acceleration
BC	uniform deceleration
CD	uniform speed (no acceleration)
DE	decreasing acceleration

EF   constant speed (no acceleration)
FG   decreasing deceleration
GH   uniform acceleration
HI   increasing deceleration

The total distance travelled between B and D is given by the total area under this part of the graph:

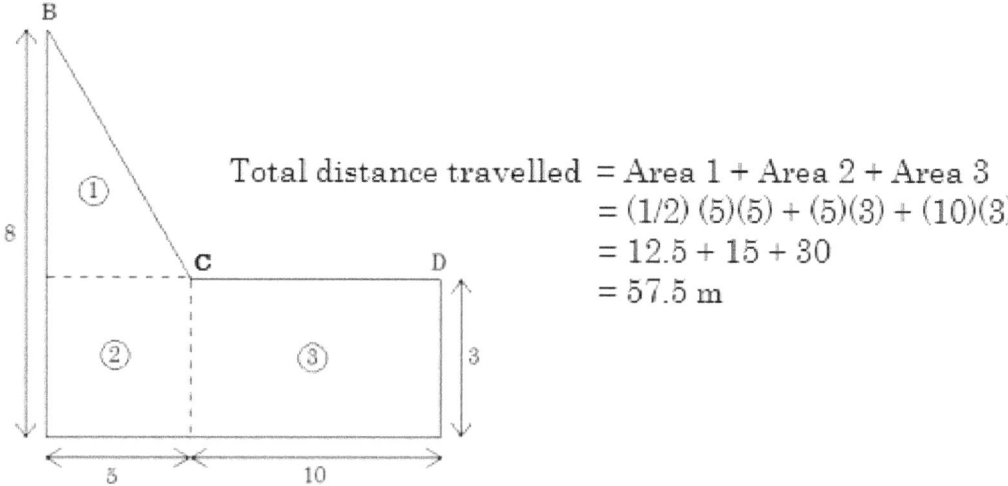

Total distance travelled = Area 1 + Area 2 + Area 3
$$= (1/2)\,(5)(5) + (5)(3) + (10)(3)$$
$$= 12.5 + 15 + 30$$
$$= 57.5\ m$$

The acceleration between B and C = (3-8)/5 = -1 m/s^2. This is a deceleration of 1 m/s^2.

To plot a distance-time graph a series of measurements are taken of the distance covered by an object at regular time intervals. These are then plotted and a line drawn through the points to produce the graph, as follows.

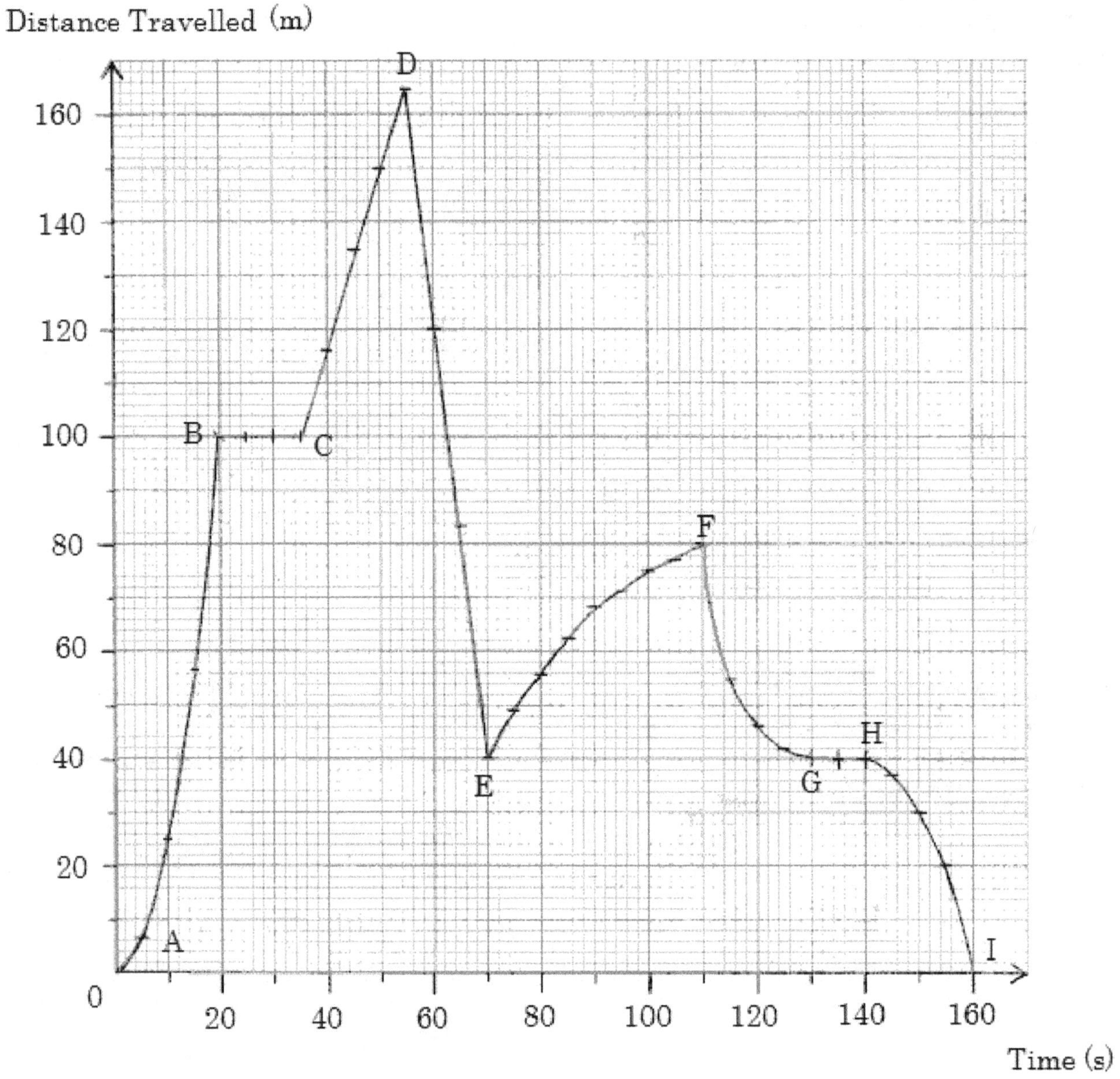

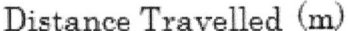

Distance Travelled (m)

Figure 2.3

A distance-time graph constructed according to the table beneath, showing the different types of motion possible between lettered regions.

Time (s)	0	5	10	15	20	25	30	35	40	45	50	55	60	65	70	75	80
Dist. Trvd. (m)	0	6	25	56	100	100	100	100	116	135	150	165	120	83	40	49	56

Time (s)	85	90	95	100	105	110	115	120	125	130	135	140	145	150	155	160
Dist. Trvd. (m)	62	67	71	75	77	80	55	46	42	40	40	40	37	30	20	0

The interpretation of different regions of the graph is as follows.

AB   uniform acceleration (increasing speed)
BC   at rest

CD   uniform speed (no acceleration)
DE   uniform speed (reverse direction to CD)
EF   uniform deceleration (decreasing speed)
FG   uniform deceleration (reverse direction to AB)
GH   at rest
HI   uniform acceleration (reverse direction to EF)

A uniform acceleration (or deceleration) has an unchanging gradient (a straight line) on a speed-time graph, and a uniformly changing gradient (a curve) on a distance-time graph. It means that an object's velocity is changing at a constant rate. Since acceleration is a vector quantity, a value can only be given where the acceleration is uniform. In the case of a non-uniform acceleration (or deceleration) over a time period, such as the region EFG of the last graph, no overall acceleration value can be given; i.e. such an average value is meaningless because the acceleration vector has changed at F.

The only true measure of acceleration is an instantaneous one, in which the change in velocity of a body is measured over an infinitesimally small timespan. Actual acceleration of a body can only be measured in this way. Even then, it is only ever an extremely precise approximation. When speaking of acceleration in the real world, what is really meant is the average acceleration over a certain timeframe. A good example of this is the standard 0-60 mph, or 0-100 km/h (0-62 mph) car acceleration test. This measures the performance of a car by recording the time taken to reach 60 or 62 mph from a standstill.

The change in velocity in the numerator of the formula for acceleration is simply the final velocity minus the initial velocity.

The conventional symbols used to represent acceleration, final velocity, initial velocity and time taken are **a, v, u** and **t** respectively, resulting in equation (1) being rewritten as equation (2):

$$a = \frac{v - u}{t} \qquad\qquad (2.3)$$

This equation applies in cases where the direction of motion does not change, e.g. accelerating in a straight line in a car on the motorway:

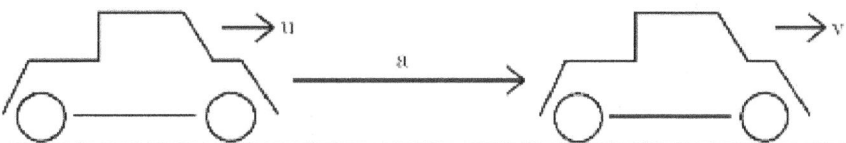

However, acceleration measures the change in velocity divided by the time taken. But velocity has a speed component and a direction component. In the example above the speed changed but the direction remained constant. It may also happen that the direction changes but the speed remains constant. This would happen if, for example, a car drove in a circle at a set speed:

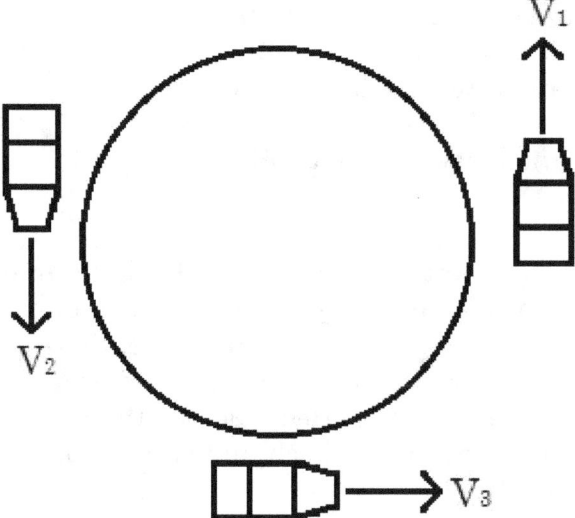

V_1, V_2 and V_3 may all have the same numerical value, i.e. the car's speed is unchanged, but they all point in a different direction. For a car to remain at constant velocity the direction as well as speed of the vehicle must be the same. Otherwise it is accelerating. This type of acceleration is called angular (or centripetal) acceleration. The reactive force to this acceleration on the car is commonly called 'centrifugal force'.

2.3 Free-Fall

An object in free-fall has two forces acting upon it; a resistive force - air resistance, and an accelerative force - gravitational pull of the earth. In an object that is falling freely, gravity is not being resisted and so the object is weightless because gravity is not felt.

All objects fall at the same rate in a vacuum, regardless of shape or weight. Even in normal atmospheric conditions, two objects of the same rigid shape that free fall while orientated in an identical fashion travel at the same velocity, even if one object is much heavier than the other one.

As the velocity of an object dropped from rest and falling through the atmosphere increases, so does air resistance, until the weight and resistive forces balance. The object can accelerate no more and has reached its terminal velocity. At terminal velocity, the forces acting on the falling object are balanced.

An object falling towards a body in a vacuum, such as the Moon in space, will continue to accelerate until it reaches and collides with the body, because the accelerative force of gravitational pull is unresisted, for example an object dropped from height over the Moon.

Section 3

Dynamics

3.1 Balanced & Unbalanced Forces

A force is a push or a pull which one object applies to another. Newton's Third Law states that:

If body A exerts a force on body B, then body B exerts an equal but opposite force on body A.

E.G.   A paperweight resting on a table pushes down on the table with its weight W, a force that is equal and opposite to the upward force, or reaction R, of the table on the paperweight.

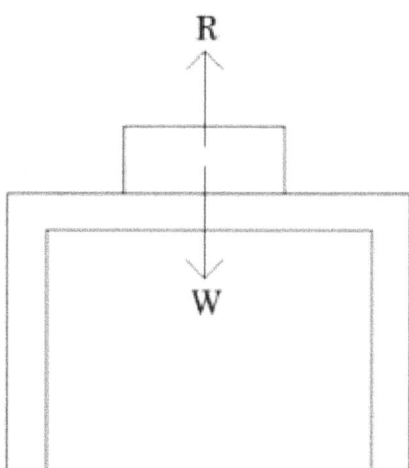

The forces on an object are either balanced or unbalanced. If the forces are balanced, then the object is either stationary or moving at constant speed. An object in which the forces are unbalanced will either be accelerating or decelerating. In the paperweight example above, the forces are balanced and the paperweight is stationary. Suppose a skydiver, having opened their parachute, is falling to earth at a steady speed of 3 m/s.

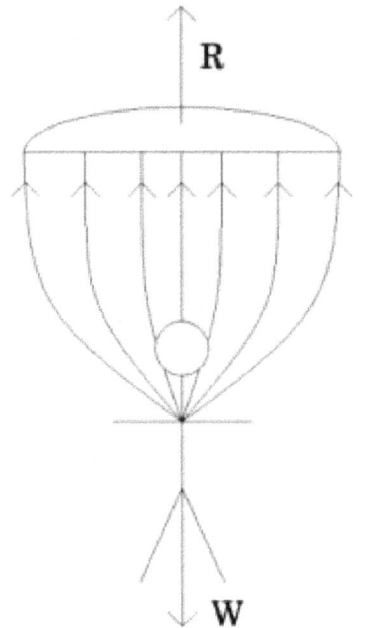

The downwards force on the parachutist, their weight, balances the upward force, air resistance or drag, exactly. This balance is designed to occur at a safe descent speed of 3 m/s. Now consider the forces acting on the skydiver at the moment they exit the aeroplane. The downward force of their weight is initially unopposed, at the start of the jump. The skydiver begins to accelerate downwards, their speed increasing all the time. But the rate at which their speed increases begins to slow, i.e. the acceleration decreases, as air resistance becomes greater and greater. Eventually, when the skydiver reaches terminal speed, the forces balance. Then when the parachute opens the forces are unbalanced again.There is still a net downwards force, but it is now much smaller, since it is balanced out by a much greater air resistance force, acting upwards. As their speed reduces so does the upwards drag force, until it eventually equals their weight, at which point they reach a new, lower terminal velocity safe for landing.

Forces cause objects to accelerate or decelerate. When objects move at constant velocity, the forces are balanced and the resultant, or net force, is zero. Forces can also prevent objects from moving; first they have to be overcome. For example, when trying to move a stationary car on a level road, the frictional force of the tyres with the road must be exceeded before the car can move.

Similarly, turning the steering wheel when the car is moving turns the car, changing its direction. The frictional force acting between the tyres and the road allows the car to turn a corner while moving without losing control. Newton's Second Law states that *a force applied to a mass produces an acceleration* according to:

$$F = m \times a \tag{3.1}$$

Force = mass x acceleration

3.2 Friction

Friction is a type of force that opposes the movement of an object at the boundary or interface where one object contacts another, usually where an object contacts a surface.

Friction is a reaction force, which means it acts to oppose movement. The level of friction between an object and a surface depends upon the properties of each. For example, pushing a flat-bottomed chest of drawers across a carpeted room is difficult because there is a high level of friction between the two surfaces which opposes movement. Pushing the same chest of drawers across a smooth marbled floor would be much easier, because these two surfaces interact less strongly.

When an object is resting on a surface, the sideways or horizontal frictional force acting on it is zero. If the object is now pushed with a gradually increasing force, the frictional force increases in proportion to the pushing force, until it reaches its limit, a maximum value where the friction can no longer increase in response to the increasing pushing force and act to oppose it. Up until this point, the object is stationary. Once this point is reached, the object begins to move in the direction of the pushing force. If, once the object is moving the force is now removed, then the object will eventually stop as the frictional force acting to slow the object down is now unopposed. Friction plays an especially important role in the safe operation of motor vehicles.

A vehicle's brakes operate because of friction between the wheel discs and brake pads. When the driver brakes, force is transmitted hydraulically from the brake pedal to the caliper, causing friction as the pads grip the disc, which slows the wheel down, and the car also.

A number of factors influence a vehicle's stopping distance. Stopping distance is the sum of the thinking distance and the braking distance.

Thinking distance is the distance travelled by the vehicle in the time it takes for the driver to react. It is affected primarily by speed and driver alertness (i.e. excess speed means a greater distance travelled in the thinking time taken to react; tiredness, alcohol and medication all slow reaction times).

Braking distance is the distance travelled by the vehicle with the brakes applied. In modern vehicles, braking distance is most affected by:

- speed

- road conditions

- condition of tyres

- condition of brakes

The braking distance of a vehicle is governed by the rate of conversion of the vehicle's kinetic energy into work done in bringing the vehicle to a stop. **Braking distance is therefore proportional to the square of the speed** according to the formula for kinetic energy. Hence doubling a vehicle's speed lengthens its braking distance by a factor of four.

In poor road conditions, the friction between the tyres and the road is reduced. When the roads are wet or icy, they become more slippery, and drivers are at greater risk of losing control of their vehicle. A car travelling at speed that hits an ice patch can lose traction and begin to slide; often this is made worse if the brakes are applied because it causes a skid and a loss of steering. A similar effect can occur with standing water, this is called aquaplaning.

Essentially the same thing happens in all these situations (skidding, sliding, aquaplaning) - instead of the wheel rolling as it does normally, it loses frictional road contact and becomes stationary as the car continues to move forward.

In wet conditions, the road-worthiness of the tyres influences braking distance significantly. The grooves in a tyre, its tread, are responsible for allowing water to flow safely underneath the tyre as it rolls forward. This prevents water from accumulating, which could cause aquaplaning. A worn tread with shallow grooves is less effective at doing its job properly and increases the likelihood of aquaplaning, or skidding under braking in wet conditions. The condition of a vehicle's brakes also affects its braking capability, worn brake pads or a thinned wheel disc mean that less force can be exerted to retard the wheel and it takes longer to stop.

3.3 Circular Motion

Newton's First Law states that:

An object stays at rest or moves in a straight line at constant speed unless acted upon by an external force.

Therefore an object moving at constant speed in a circular path must be acted upon by an external force, otherwise it would travel in a straight line. This force is called a centripetal force, and always acts perpendicularly (at right angles) to the direction of movement of the object. If centripetal force did not act perpendicularly to the object at all times, then it would cause the object to speed up or slow down.

Consider a cylist moving at constant speed around a circular track.

The force the cyclist produces by pedalling may be resolved into two components, F_1 and F_2. F_1 is the part of the force used to overcome air resistance and the friction of the tyres with the road, F_2 is the frictional force between the road and tyres that acts sideways towards the centre of the circle. It keeps the bicycle moving around in a direction tangential to the circle. Without it, the bicycle would just move in a straight line.

Other examples of centripetal force include:

- the electrostatic force acting on an electron in an atom, which keeps negatively charged electrons orbiting around a positively charged nucleus.

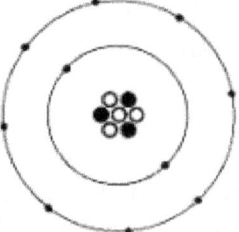

- the gravitational force on a satellite (man-made or natural) orbiting a planet, which causes the planet's mass to attract the mass of the satellite by gravity.

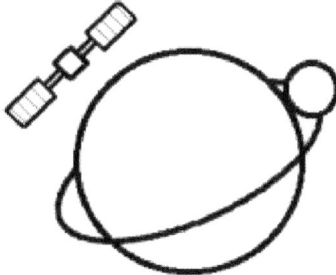

- the gravitational force of the Sun, due to its mass, which attracts the smaller masses of the planets and causes near circular (elliptical) motion of their orbits.

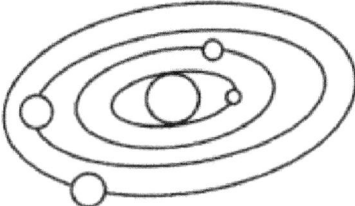

Section 4

Mass, Weight & Density

4.1 Mass & Weight

Mass is a measure of the amount of substance, or matter ('stuff'), in a body. It is an SI quantity and is measured in kg.

In accordance with Newton's First Law, the mass of an object resists change from its state of rest or motion. The object has a 'reluctance' to begin moving from rest, or stop moving once started. This is called inertia. The mass of an object is a measure of its inertia, so a large mass has a large inertia and a small mass a small inertia.

Gravitational Fields

A large body such as a planet produces an invisible gravitational field. A gravitational field is a region in which a mass experiences a force due to gravitational attraction. Newton's second law has been stated as F=ma. In this equation, a mass that is accelerated does so because of a force acting on it. If this acceleration is directed towards the earth as a result of earth's gravitational field, the the force F resulting from this is termed the weight W. Rearranging F=ma to make the acceleration a the subject of the equation gives:

$$a = \frac{F}{m} \qquad (4.1)$$

When force F is replaced by weight W in this equation to signify the weight force of an object in a gravitational field, acceleration a is conventionally represented by the letter g, to signify gravitational field strength:

$$g = \frac{W}{m} \qquad (4.2)$$

Equation (2) may alternatively be stated as W=mg.

Weight, as a force, is measured in Newtons (N) and mass is measured in kilograms (kg), so gravitational field strength is measured in Newtons per kilogram (N/kg). The earth's gravitational field strength is 9.8 N/kg.

A chemical, or beam balance is used to measure mass. It compares unknown masses with known masses. See figure 4.1 for an illustration. Delicate pivot points at A, B and C allow the scales to be finely balanced, which produces an accurate measurement.

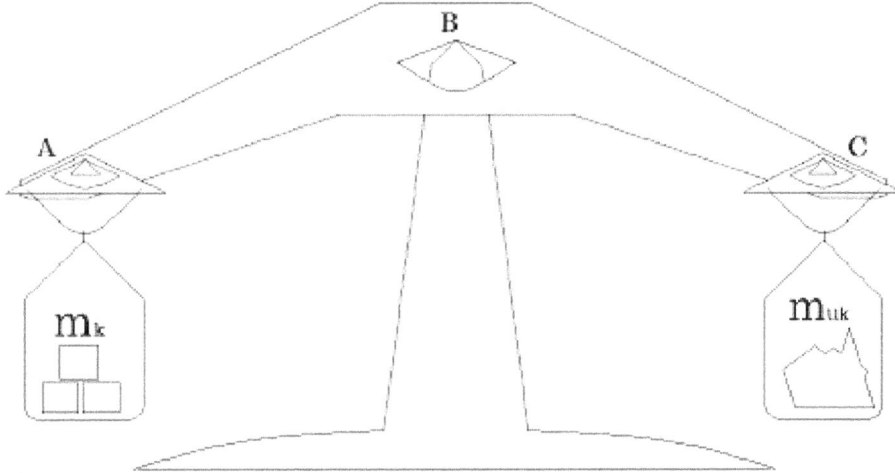

Figure 4.1

A chemical balance with an unknown weight m_{uk} determined from the sum of known weights m_k.

The unknown mass which is to be determined is placed in one of the scale-pans. Masses of known value are then placed in the other scale-pan, in successively smaller units, until the centre dial is aligned with zero on the scale, indicating that the balance point has been reached. The pivots at A and C ensure that the force due to the weight of the masses in either pan acts vertically downwards exactly, as if an imaginary line were extended from the pivot points to pass directly through the centres of gravity of the masses beneath.

When the beam is balanced the unknown mass equals exactly the sum of the known masses that counterbalance it. Gravity exerts an equal force on either pan, and can be cancelled out of the equation. Thus:

weight: $m_{uk} \cdot g = m_k \cdot g$ at balance
leading to $m_{uk} \cdot g = m_k \cdot g$
i.e. $m_{uk} = m_k$

A spring balance is used to measure weight. Unlike a chemical balance, which gives a comparative measure of mass, a spring balance provides an absolute measurement of weight. A bathroom scales works in this way. In experimentation, the balance is usually held by a clamp, which is attached to a retort stand and suspends the balance above the workbench. A calibrated spring whose extension properties are known resists a weight hooked to it, producing an extension on a uniform scale. A 'uniform' scale is one in which equal divisions indicate equal weight over the whole length.

The weight readings obtained using a spring balance will vary slightly depending on where they are made on Earth. At the equator, the radius of the earth is slightly bulged, meaning it is a tiny bit further from the earth's centre than on average, so gravity here is a bit weaker than in most other places. Consequently an object weighed here will record a reading a bit lower than found almost everywhere else. At the poles however, the same object will record a higher weight reading than in other locations, because the earth is slightly flattened here, which means the surface is closer to the centre of earth on average. It therefore experiences a higher gravitational force than elsewhere.

For this reason, most space rocket launches occur at tropical latitudes near to the equator, where earth's gravitational pull is at its weakest due to the bulging of this region.

4.2 Density & Volume

Density is an object's mass per unit volume.

$$\text{Density} = \frac{\text{mass}}{\text{volume}} \tag{4.3}$$

Density is measured in kg/m^3. Water at a temperature of $4°$ has a volume of 1000 kg/m^3. It is relatively straightforward to find the density of a liquid. A measuring cylinder is placed on an electronic balance (digital scales) and the readout is tared (set to zero). The liquid is poured into the cylinder and it's mass is determined from the readout. The volume of liquid in the cylinder is then read off the measuring cylinder's scale. The mass and volume figures are put into formula (3) and the density is calculated.

Determining a solid's density is a little more difficult. The mass is found in the same way, by using an electronic or chemical balance, but finding the volume is more involved.

For a regularly shaped, homogeneous (the same material and properties throughout) solid such as a cuboid or sphere, the dimensions are found in terms of width, length, height or diameter and these are used to calculate its volume. Its density is then found by inserting these figures into the formula, as before.

The volume of an irregularly shaped, homogeneous solid is difficult to determine accurately by direct measurement without expensive equipment utilising laser technology. Fortunately though, an indirect but effective method exists, which is suitable for solids that sink in water. It is therefore applicable to many materials, including stone, ceramics and most metals. The irregular solid is immersed in a measuring cylinder pre-filled with a known volume of water. Upon sinking, it displaces its own volume of water and the level of the water rises. The new volume reading is noted and the original reading, without the solid, is subtracted from it. This leaves a difference, which is the volume of water displaced by the solid. This equals the volume of the irregular solid. It's mass is then found using the density formula. This measurement technique is known as *volume by displacement*.

For objects which are larger and cannot fit in a measuring cylinder, a Eureka can may be used, as illustrated below. The water level is filled right up to the short spout. The solid is then immersed and the amount of water displaced into the beaker gives the volume of the solid.

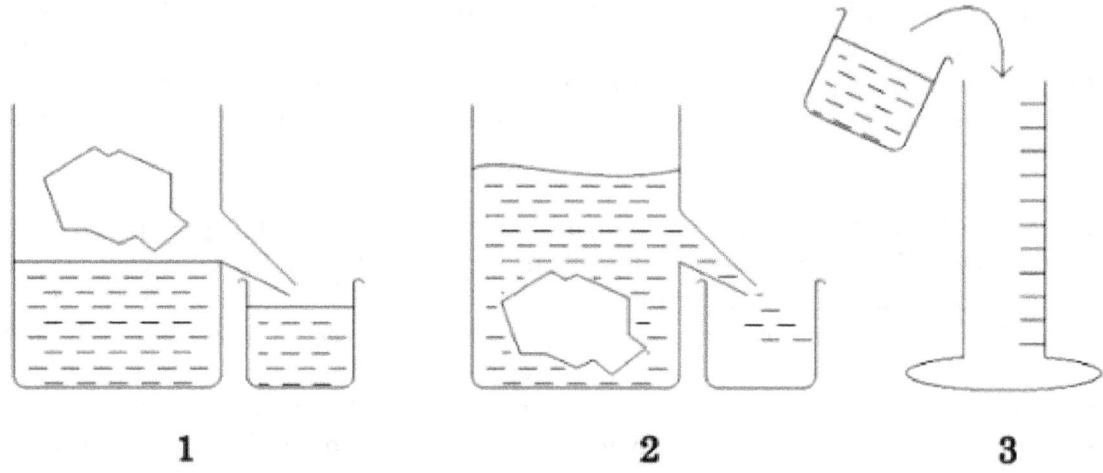

1   **2**   **3**

Figure 4.2

An illustration of the volume by displacement method for finding the volume of an irregular solid.

Section 5

Turning Effect of Forces

5.1 Moments

A moment, or torque, is the turning effect of a force about an axis orthogonal (perpendicular) to its plane of action. For objects which are pivoted about an axis, the moment of a force, M, is given by:

$$M = F \, x \, d \tag{5.1}$$

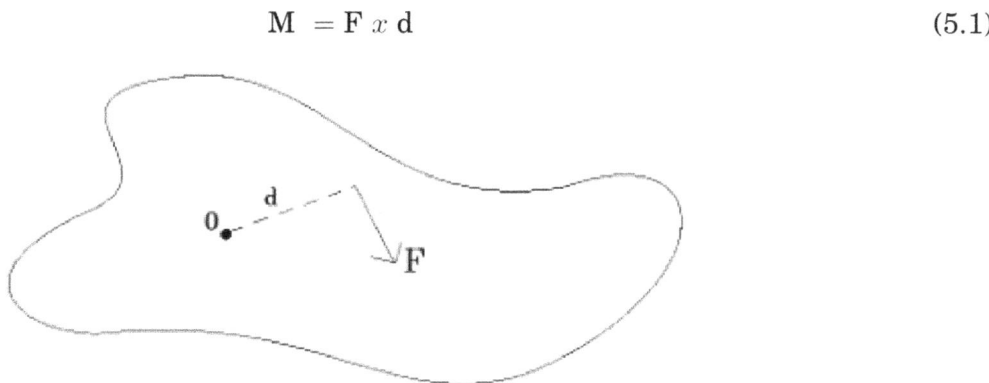

$$\text{Moment of a force} = \text{Force } x \text{ Perpendicular distance from pivot} \tag{5.2}$$

The moment of a force acting far away from a pivot is greater than if the same force acted close to the pivot. A large force also produces a greater moment than a smaller one. On a hinged door, for example, the knob or handle is usually at the opposite end of the door to the hinge (pivot), so that it provides as much leverage as possible. This means that only a small force is required to open or close the door. If however, the knob was placed close to the hinge, it would require a very large force to open and close the door and would be impractical. A spanner makes use of the same principle in order to tighten or loosen a nut on a bolt. The maximum force which can be exerted by a person with a spanner cannot be increased. However, the arm of the spanner can be made longer, thus increasing the distance between the pivotal axis, at the nut's thread, where the force acts, and where it is applied, near the end of the spanner's arm. The applied moment is now larger than before, even though the force used remains the same, so the turning effect is greater.

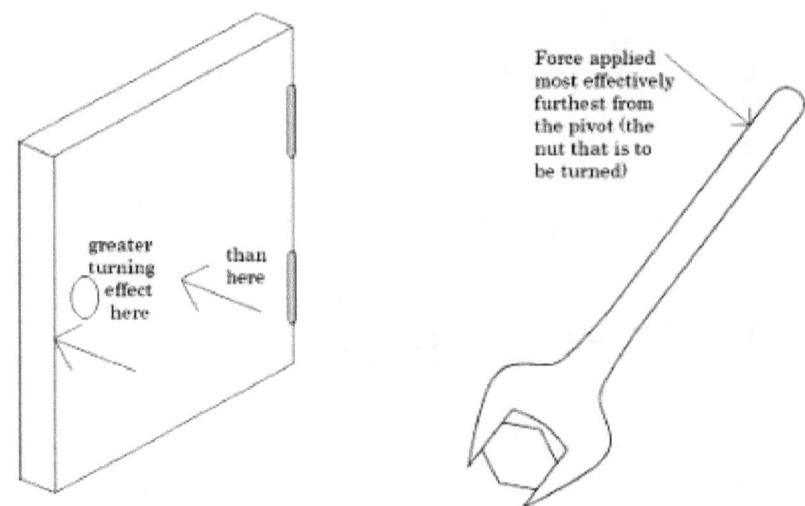

Figure 5.1

An applied force provides the greatest turning effect, or moment, furthest from the pivot.

E.G.   A moment of 15 Nm is required to open a door at its outside edge. If the door is 0.6 m
wide, what is the force required?

$$M = F \times d$$
$$15 = F \times 0.6$$
$$F = \frac{15}{0.6} = 25 \ N$$

(5.3)

An object which is in equilibrium has no unbalanced moments or forces acting on it. The
principle of moments for an object in equilibrium states that:

Sum of clockwise (c/w) moments about any point =
                    Sum of anticlockwise (a/c) moments about same point

The following three examples illustrate these concepts.
(i)   A uniform beam of mass 8 kg has 3 forces plus its weight acting on it. The beam is in
equilibrium. One of the forces, F, is unknown; determine its value (Take gravitational
field strength, g to be 10 N/kg).

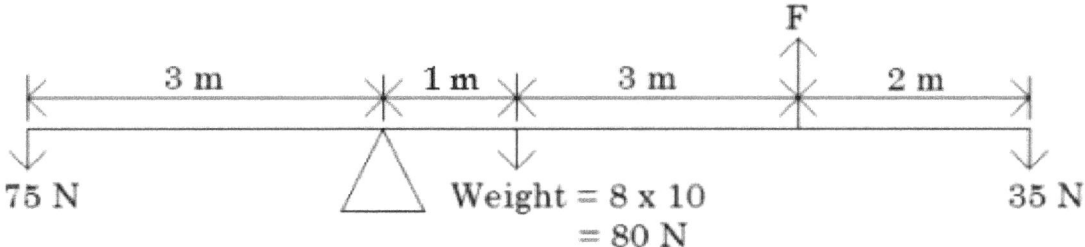

The pivot is located 3 m from the left edge of the beam. The beam is uniform in cross-section
so its weight acts at its centre of mass, in the centre of the beam, 4 m from either edge. Taking
moments about the pivot, the *principle of moments* applies:

sum of clockwise moments = sum of anticlockwise moments

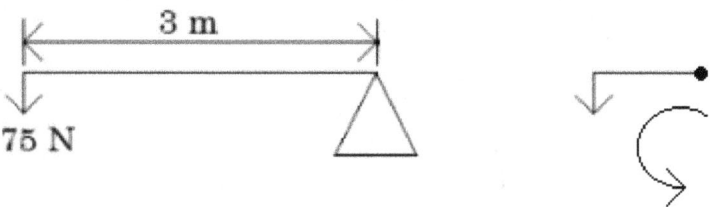

Anti-Clockwise Moment = 75 x 3 = 225 Nm

Clockwise Moment = 80 x 1 = 80 Nm

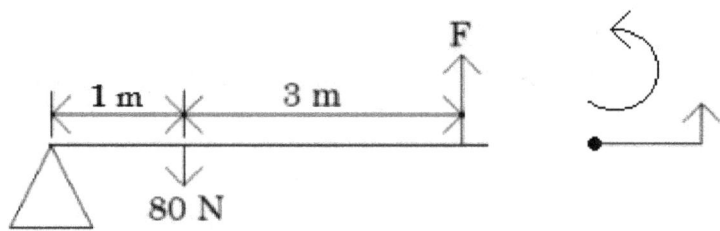

Anti-Clockwise Moment = F x 4 = 4F

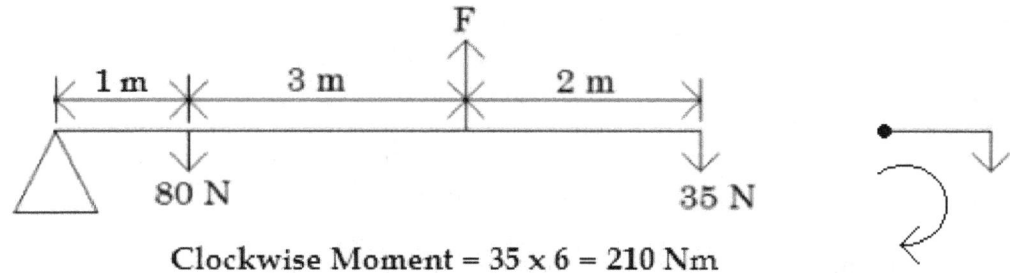

Clockwise Moment = 35 x 6 = 210 Nm

Sum of Clockwise Moments = Sum of Anticlockwise Moments

80 + 210 = 225 + 4F

290 = 225 + 4F

290 - 225 = 4F

65 = 4F

F = 65/4 = 16.25 N

(ii)   A zero-weight beam has 5 forces acting on it to keep it in equilibrium. Two of the forces are unknown. Find each given that F_2 is three times the size of F_1.

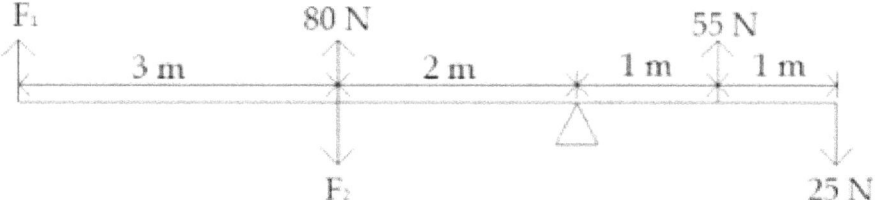

Taking Moments about the pivot:

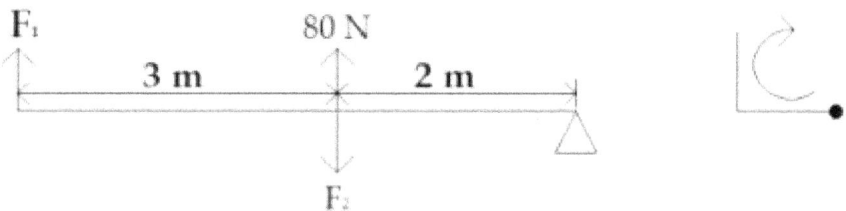

Clockwise Moment $= F_1 \times 5 = 5 F_1$

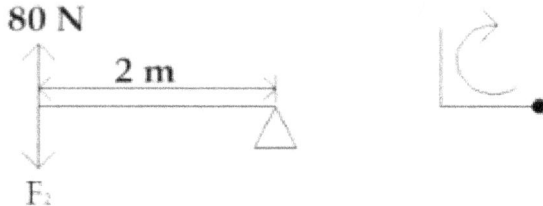

Clockwise Moment $= 80 \times 2 = 160 \ Nm$

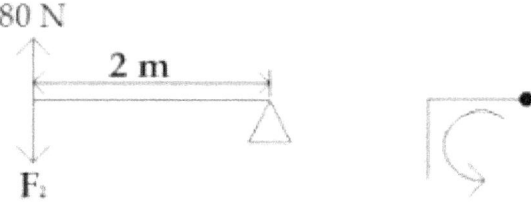

Anticlockwise Moment $= F_2 \times 2 = 2 F_2$

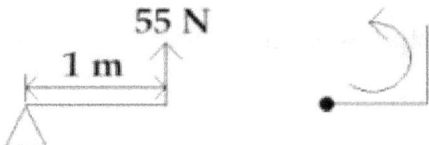

Anticlockwise Moment = 55 x 1 = 55 Nm

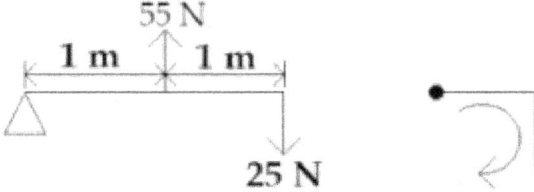

Clockwise Moment = 25 x 2 = 50 Nm

Sum of Clockwise Moments = Sum of Anticlockwise Moments

$$5 F_1 + 160 + 50 = 2 F_2 + 55$$
$$5 F_1 + 210 = 2 F_2 + 55$$
$$5 F_1 + 210 - 55 = 2 F_2$$
$$5 F_1 + 155 = 2 F_2 \quad ... [1]$$

Knowing that: $F_2 = 3 F_1$ ... [2]

Substituting from equation [2] into equation [1] for F_2 gives:

$$5 F_1 + 155 = 2 (3 F_1)$$
$$5 F_1 + 155 = 6 F_1$$
$$F_1 = 155 \text{ N}$$
$$F_2 = 3 F_1 = 3 \times 155 = 465 \text{ N}$$

(iii) A see-saw pivoted at its mid-point has 3 children seated on it. A heavy bag placed somewhere along its length keeps it in equilibrium. Given that the see-saw's mass is 15 kg and acts through the pivot, and that the children's weights are as shown, find the location of the bag given that it weighs 10 kg.

$$400 \times 4 + 350 \times 3 = 100x + 600(4)$$
$$1600 + 1050 = 100x + 2400$$
$$2650 = 100x + 2400$$
$$2650 - 2400 = 100x$$
$$250 = 100x$$
$$250/100 = x$$
$$x = 2.5 \text{ m}$$

The bag is 2.5 m from the pivot on the right hand side. Note that the see-saw's weight is not relevant to the question, because its weight acts through the pivot.

The principle of moments may be verified experimentally using a beam balance. The beam resembles a ruler that is drilled at even intervals along its length, coinciding with a marked scale. At its midpoint the beam is balanced on a knife-like edge, the pivot. The beam is usually held in place with a boss and clamp arrangement, attached to a retort stand. The principle of moments is investigated by adding hooked mass holders to either side of the beam. These are moved from one hole to another and/or slotted masses added to achieve a balance point by trial and error. Once a balance point has been achieved, the values and positions of all masses are recorded in a table. Different configurations of masses are balanced in the same way and recorded also. For each entry in the table, the sum of the clockwise and anticlockwise moments are compared, as before. These should be found to be equal, thus providing verification for the principle of moments.

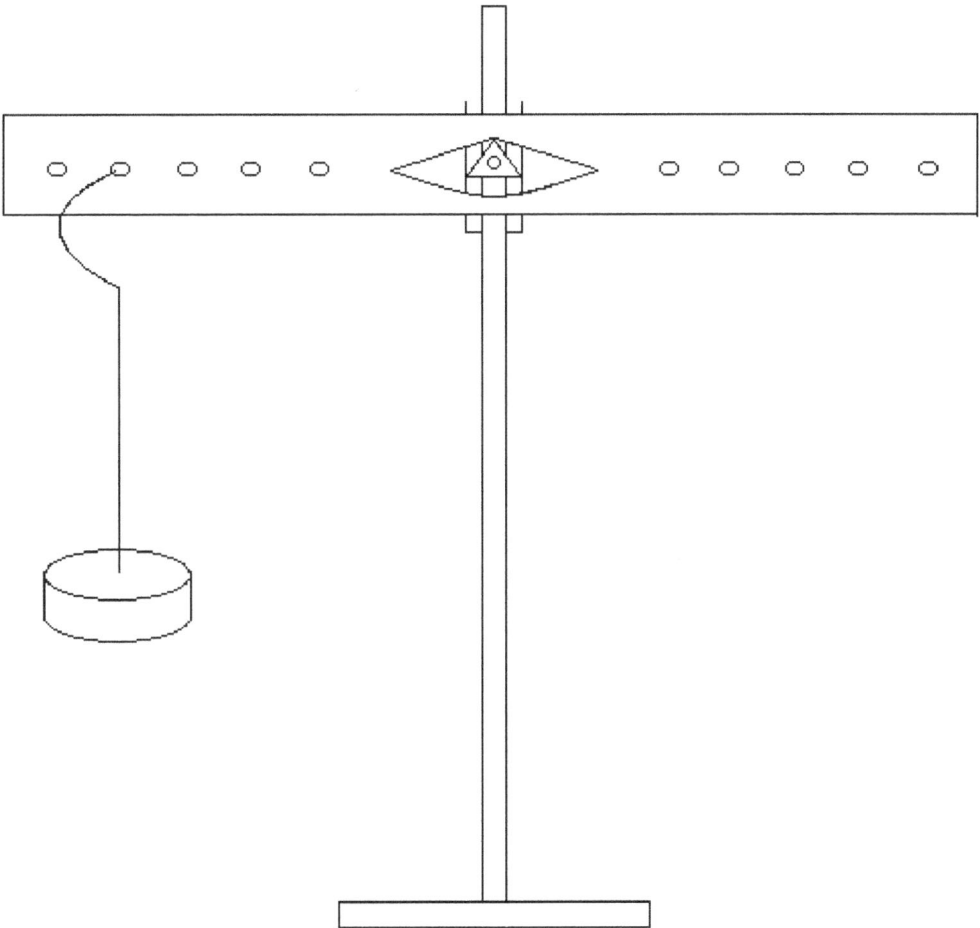

5.2   Centre of Mass

The centre of mass is the single point where the entire weight of a body seems to act. A plane lamina is a two-dimensional object (it has zero thickness). This means that its geometric centre coincides with its centre of mass, as its mass is uniformly distributed. To determine the centre of mass of a plane lamina, the geometric centre of its composite shape(s) must be found. A complicated plane lamina is broken up into its constituent shapes (square, circle, rectangle, etc), their individual centres of area are found, and the results added together to give a value for the whole shape. This is its centre of mass. See the following example.

Find the centre of mass of the following plane lamina.

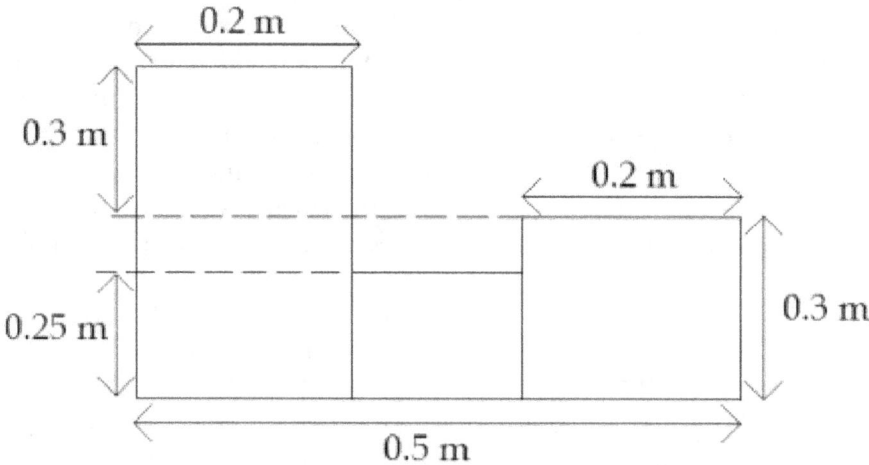

First, the shape is broken up into three rectangular pieces. A cross marks the centre of mass of each:

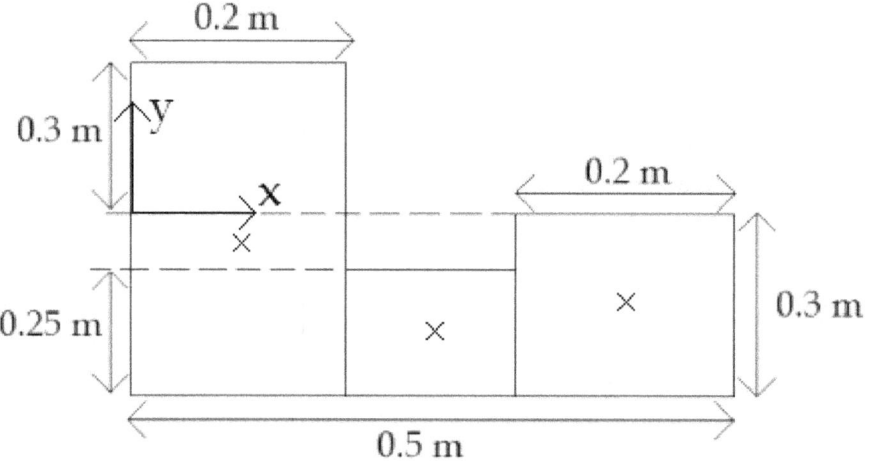

A suitable datum point is chosen where all distances are referenced to and from. x and y are positive directions from the datum point. A table is drawn up which records the horizontal and vertical distances of each shape's centre of mass from the datum:

Table 5.1: Centre of Mass Tabulations

Shape	Area (m^2)	x dist to CoM (m)	Area * x-dist (m^3)	y-dist to CoM (m)	Area * y-dist (m^3)
1	0.12	0.1	0.012	0	0
2	0.025	0.3	0.0075	-0.125	-0.003125
3	0.06	0.4	0.024	-0.1	-0.006
Sum:	0.205	Sum:	0.0435	Sum:	-0.009125

It is as if a single area of 0.205 m^2 acts at one point with coordinates (x_m, y_m). When the whole area is multiplied by the x-coordinate of its centre of mass, x_m, it produces a value equal to the sum of all the individual areas multipled by the x-coordinates of their centres of mass. Thus

$$0.0435 = 0.205 * x_m \tag{5.4}$$

i.e.

$$x_m = \frac{0.0435}{0.205} = 0.212 \text{ m (2 d.p.)} \tag{5.5}$$

Similarly, the whole area multiplied by the y co-ordinate of its centre of mass, y_m, produces a value equal to the sum of all the individual areas multiplied by the y-coordinates of their centres of mass. Thus:

$$-0.009125 = 0.205 * y_m \tag{5.6}$$

i.e.

$$y_m = \frac{-0.009125}{0.205} = \text{-0.0445 m (4 d.p.)} \tag{5.7}$$

5.3 Stability

Stability is a measure of an object's resistance to toppling if displaced by a force from its equilibrium position. The stability of an object depends on the position of its centre of mass and the size of its base area. A stable object invariably has a low centre of mass and a wide base.

Objects are classified in three ways according to how stable they are:

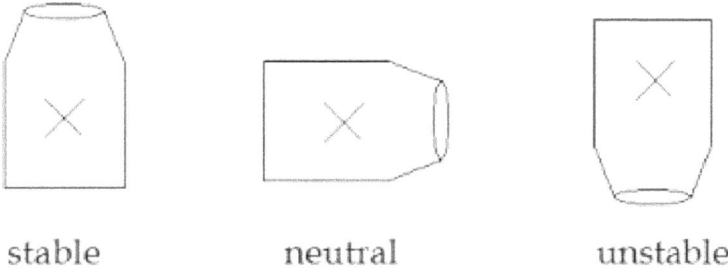

stable   neutral   unstable

These states are more properly called equilibrium positions.

In stable equilibrium, the object returns to its original position when tipped slightly by a force. The milk bottle in the example has a low centre of mass and a fairly wide base.

In neutral equilibrium, the object remains in the same position when a force tilts. The milk bottle in neutral equilibrium has a very wide base and the lowest centre of mass. An object in unstable equilibrium moves further away from its original position when tilted by a small force. The milk bottle example has a narrow base and a high centre of mass. For these reasons, high-sided vehicles such as lorries and buses are designed so that the centre of mass is as low as possible.

All heavy items including the engine and mechanicals are mounted as low as possible to bring down the height of the centre of mass. Also, in the case of a bus, upstairs passengers are not permitted to stand while journeying, as this would raise the overall centre of mass of the bus to an unacceptable level, resulting in an increased susceptibility to toppling.

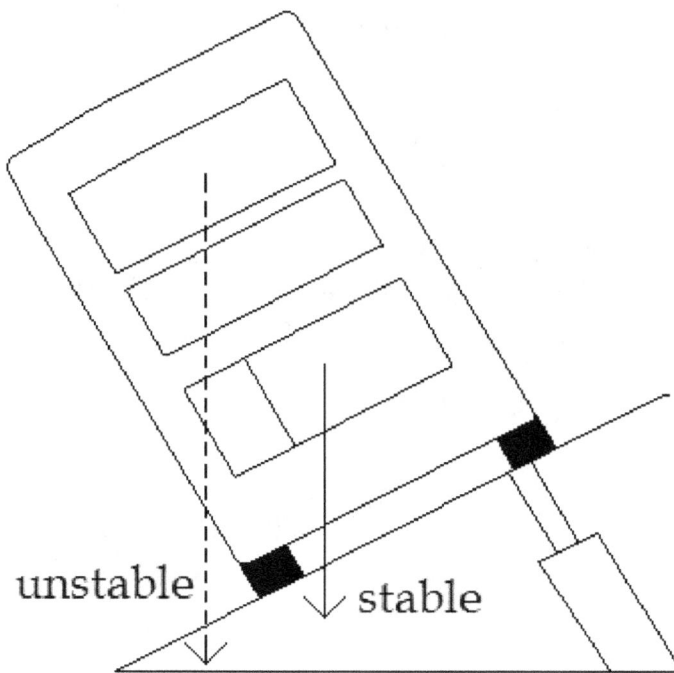

Despite being tall, a buses' centre of mass is as low as possible and its base as wide as possible, so that even if it is tipped, the centre of mass will not be over a point that falls outside the base.

The design of everyday items such as coat stands, reading lamps and wine glasses are also affected by these considerations. For this reason, all have a very wide, low and heavy base.

Section 6

Elastic Deformation

Applying a force to an object may produce a change in the object's size and shape. The applied force acts to either compress or extend the object. This produces a deformation that is either temporary or permanent. An elastic object is one which deforms reversibly under a certain force. This means that when the force is removed, it returns to its original shape as if the force had never been applied. All solid materials deform elastically under very low forces; those which continue to do so as the forces are increased greatly are said to be elastic. Materials which deform permanently by stretching under sizeable forces are said to be plastic. Most metals are elastic. Some, like tin and lead, exhibit plastic-like properties. Metals such as these are more usually referred to as being malleable or ductile than as plastic though, as this description better fits their material composition and does not lead to confusion.

Elastic materials deform temporarily in response to a force, although there is a limit to the extent of this deformation, before they lose elasticity and either break or become permanently deformed. This is called the **elastic limit**. As well as metal; glass, wood, concrete and rubber are commonly occurring elastic materials. Despite being elastic, a material's level of elasticity, or strength, may vary widely depending on the types of load being applied. Steel, for example, is very strong in both compression (pushing together) and tension (pulling apart). Concrete however, is very weak in tension, despite being strong in compression. It has a low elastic limit in tension. For this reason, concrete is often reinforced with steel (rebar), which increases its tensile strength. Materials such as steel and concrete, whose properties are of crucial importance in the safe and efficient construction of buildings and roads infrastructure, are tested extensively in engineering laboratories. This testing takes place so that the properties and behaviours of the materials are properly and thoroughly understood.

In most testing regimes, increasing loads are applied to a material sample whilst the deformation is simultaneously recorded; instruments perform measurements continuously. To measure the very small elastic deformations produced by sample specimens of engineering materials, often on a scale of thousandths of a millimetre, specialised instrumentation is used. However, the loading machines and measuring equipment involved in meaningful testing are beyond the scope of simple laboratory experimentation. Fortunately, an analogous method can be used in the laboratory to demonstrate the principle effectively. This involves an investigation into the extension-load properties of steel springs and rubber bands. Both produce large extensions for modest loads, allowing practical examination of the extension-load relationship. The experimental apparatus is set-up as in figure 6.1.

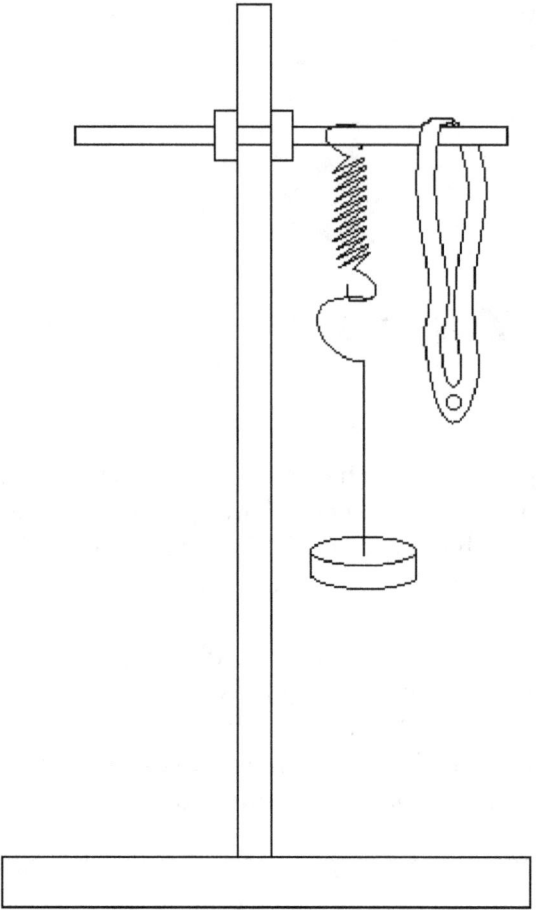

Figure 6.1

The experimental setup to measure the extensions of a rubber band and coiled steel spring, when weighted by masses

A hole is punched through the free end of the rubber band, which is reinforced with aluminium tape. The unextended lengths of the spring or rubber band are measured while they are unweighted. A mass holder is then added to each, the mass noted, and the corresponding extension observed and recorded. The procedure is continued by adding masses in 25 g denominations, each time reading the new extension and recording it. The graphs produced for each should be as follows:

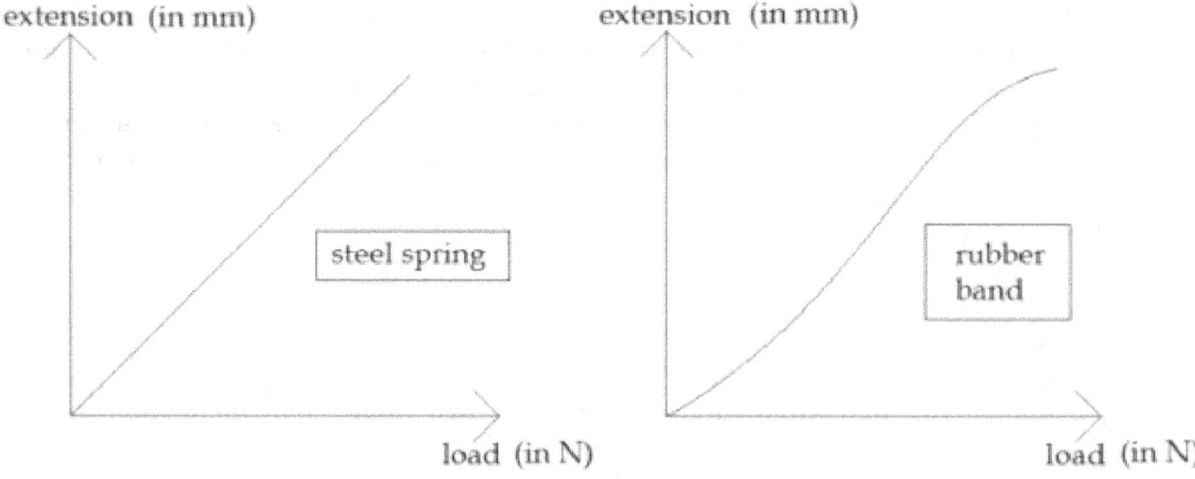

The graphs look very similar, but the plot for the rubber band is curved slightly in an s-shape, whereas the spring plot is a straight line. The extension of the steel spring is proportional to the load applied. It obeys Hooke's Law, which states that extension is proportional to load until its "limit of proportionality" is reached:

$$\text{extension } \alpha \text{ load}$$

The **limit of proportionality** is slightly less than the elastic limit, which means that a material can still return to its original shape with no permanent deformation, even though it may have passed the limit of where its extension is proportioanl to the load applied. The elastic limit is usually not much further than the limit of proportionality though; they can be thought of to approximately coincide.

Rubber does not obey Hooke's Law, at no point on the load-extension curve is there a linear relationship between the two properties.

Figure 6.2 below shows an extended version of the two previous graphs, which incorporates an unloading cycle for each. This is shown by the dashed line in either case. The steel spring is permanently deformed, it has been loaded beyond its elastic limit into the plastic zone and now exhibits a permanent extension, even after complete unloading. The contours of the curve for the rubber band have been exaggerated to make the shape more pronounced, it has not been extended beyond its elastic limit, but nevertheless displays hysteresis, a phenomenon which weakens its resistance to load due to heating upon loading. It means that as it is unloaded, it eventually recovers its original shape, but for the extension produced by a given load upon loading, the equivalent extension at the same load upon unloading is greater. The joint loading and unloading curve for rubber is known as a hysteresis loop.

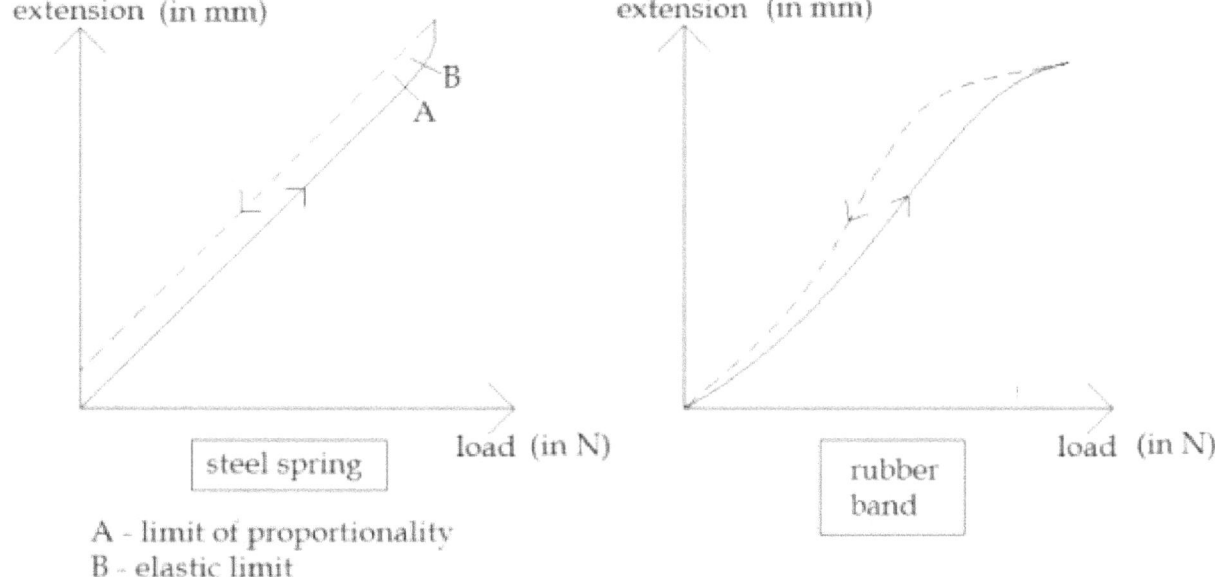

A - limit of proportionality
B - elastic limit

Figure 6.2

Loading-Unloading Curves for the steel spring and the rubber band

For an elastic solid that obeys Hooke's Law, the governing equation may be written as:

$$F = kx \qquad\qquad (6.1)$$

where F = applied force (N)
     x = extension (m)
     k = spring constant (N/m)

The following two examples demonstrate the use of this equation.

(i) A spring's unloaded length is 6 cm. When a 200g mass is added, it extends by 4 cm. When the mass is removed it returns to its original length.

(a) Find the value of the spring constant

(b) If a 350 g mass is added instead, and the spring returns to its original length as before once the mass is removed, what extension does the new mass produce? (Take acceleration due to gravity to be 10 N/kg)

(a) The spring obeys Hooke's Law.

Therefore F = kx
200g = 0.2 kg
4 cm = 0.04 m
g = 10 N/kg

By Newton's Second Law, F = ma
i.e. F = mg = (0.2)(10) = 2 N
Also, extension x = 0.04 m

Now

$$k = \frac{F}{x} \qquad (6.2)$$

giving k = 2/0.04 = 50 N/m

(b)
350 g = 0.35 kg
New Force F = mg = (0.35)(10) = 3.5 N
k = 50 N/m from part (a)
Therefore x = F/k = 3.5/50 = 0.07 m = 7 cm

(ii) All the springs shown in diagrams (a) to (c) are identical. The extension of the spring in (a) is 5 cm. Find the individual extensions of the springs in (b) and (c), determine the overall extension, and find the spring constant.

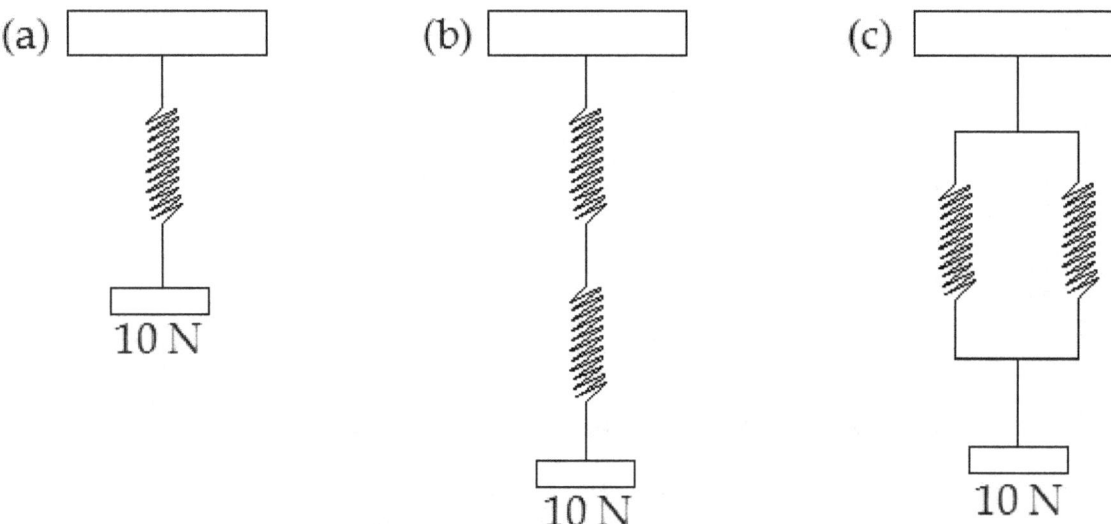

Spring constant k = F/x
Extension for spring (a), x = 0.05 m
Force on spring (a), F = 10 N
Thus Spring constant k = 10/0.05 = 200 N/m

In (b), the springs are in series. They both feel the same force. The individual spring extensions are therefore the same. Both springs extend by 5 cm. The total downward extension is therefore 10 cm. In (c), the springs are in parallel. They share the load, so each feels only half the force. Each spring extends by 2.5 cm. The total downwards extension is therefore 2.5 cm also.

Section 7

Pressure

Pressure is a measure of the distribution of force over an area.

$$\text{Pressure} = \text{Force/Area}$$

$$p = \frac{F}{A} \qquad\qquad (7.1)$$

Pressure is measured in N/m^2, or pascals (Pa).

E.G. A force of 200 N is used to hammer a nail into a block of wood. If the area of the head of the nail is $1mm^2$, what is the pressure exerted on the wood?

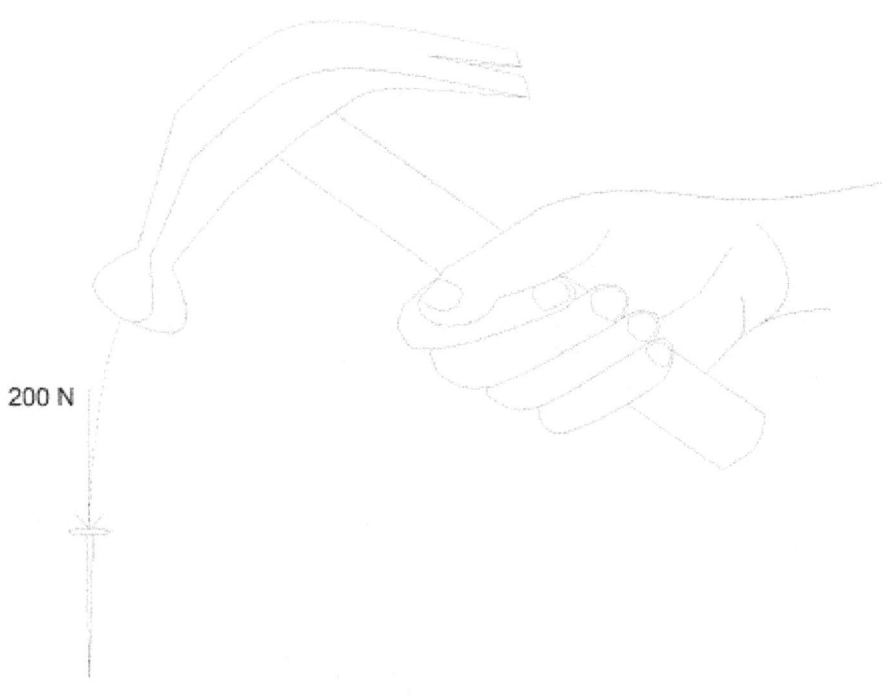

200 N

$1\ mm^2 = 1 \times (1/1000)^2 = 1 \times 10^{-6}\ m^2$
$p = F/A = 200/(1 \times 10^{-6}) = 200 \times 10^6 = 200\ MPa$

(This is equivalent to a mass of 20 000 tonnes acting over an area of 1 square metre.)

A force acting over a small area exerts a pressure much greater than the same sized force acting over a large area. For example, a military tank can cross soft, 'boggy' marshland without sinking more than a few inches, since its wide caterpillar tracks spread its mass evenly over a large area. An infantry soldier however, is unable to cross the same terrain without sinking to his knees, making forward progress very difficult. Figure 7.1 compares the pressures involved in either case.

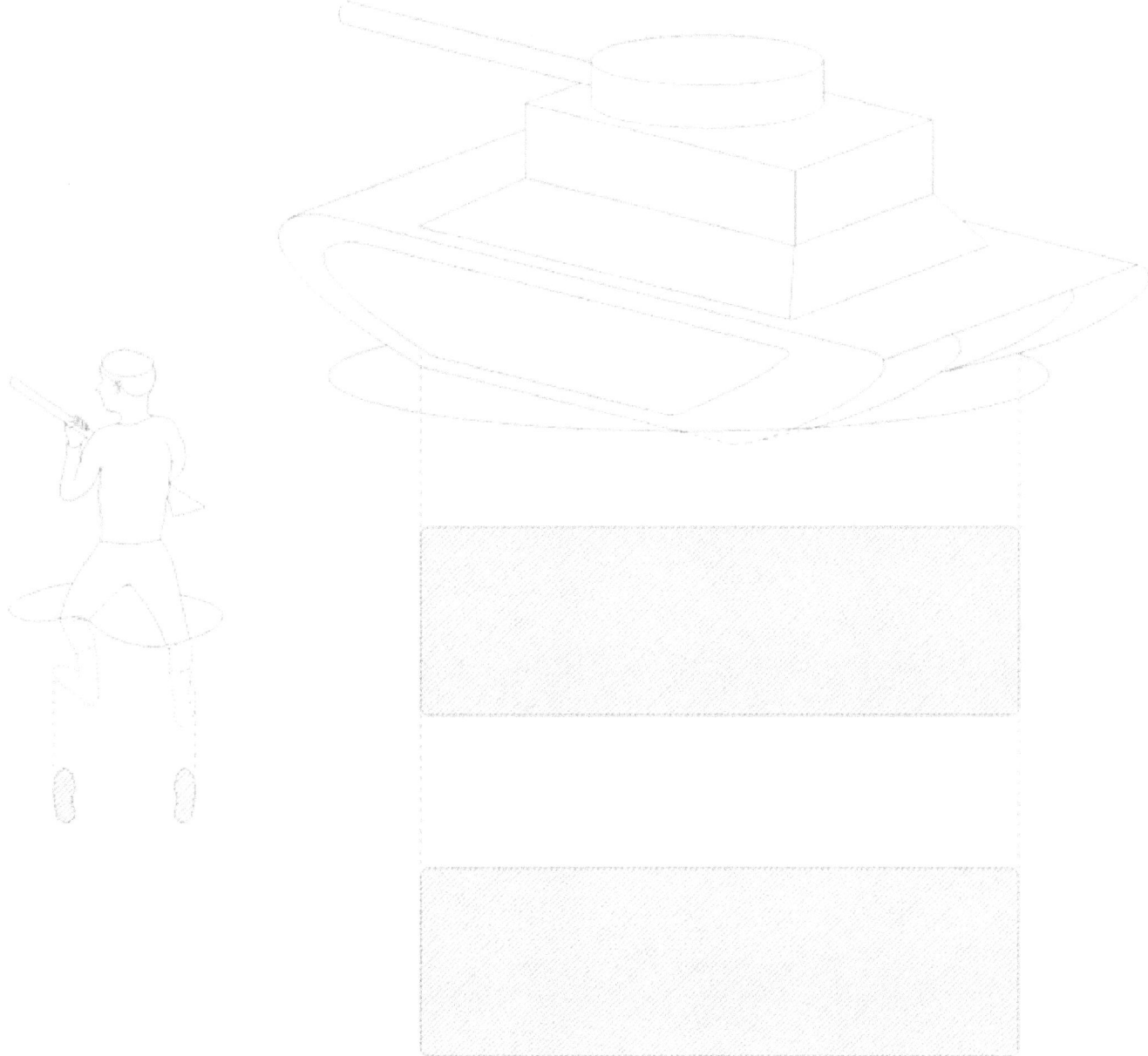

Figure 7.1

A tank sinks less deep into boggy ground than a soldier does, despite a soldier weighing much less, because the tank's weight is spread over a wide area, whereas the soldier's weight is concentrated over a small contact patch

The scale of the soldier to the tank is 2:1. The soldier's total mass including equipment is 110 kg. The sole of each of his boots has an area of 200 cm^2 = 0.02 m^2. Taking g to be 10 m/s^2, the pressure exerted by the soldier on the ground is given by:

p = F/A = (110 x 10)/(2 x 0.02) = 27,500 N/m^2 = 27.5 kN/m^2 (kPa)

The tank weighs 25 000 kg, and each track has a contact patch of 10 m^2. Thus the pressure exerted on the ground by the tank is as follows:

p = F/A = (25000 x 10)/(2 x 10) = 12,500 Pa = 12.5 kPa

The soldier sinks 18 inches (457 mm) into the ground. Given the pressures just calculated, how deep does the tank sink? Let x represent the depth in metres to which the tank sinks.

Ratio of sinking depths = Ratio of pressures exerted

457/x = 27.5/12.5

Thus x = 457 x (12.5/27.5) = 208 mm (about eight inches)

Similarly, a lady wearing very narrow 'stiletto' heels may indent soft floor coverings due to the high pressure exerted at the heel. If she wears wide-heeled shoes however, the force is spread more widely and the shoes do not leave a mark. Compare the contact area of a stiletto shoe, as shown below, with that of the soldier's boot in figure 7.1.

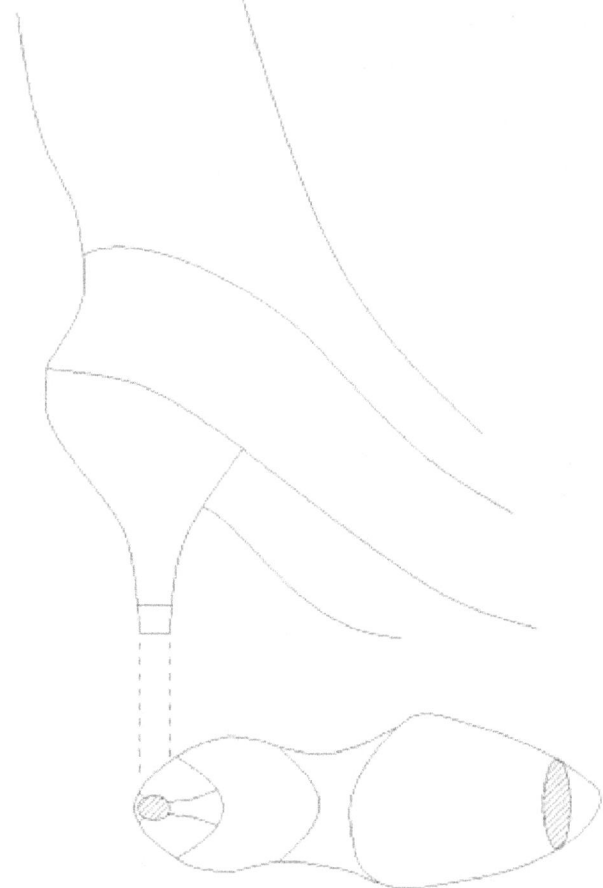

The lady weighs 550 N. Each shoe has a total contact area of 200 + 75 = 275 mm^2. The pressure is therefore:

p = F/A = 550 / (2 x 275 x (1/1000)^2) = 1 MPa, i.e. 36 times more pressure than the soldier exerts in his flat-heeled boots.

In some cases, the effect of a narrow contact area is deliberately exploited by designers in order to create high pressures there. A knife for example, cuts through butter easily because it has a very finely tapered edge; this gives it a minute contact area, translating the small force applied to a high cutting pressure. In the same way, an ice skate has a thin contact edge to produce high pressures at the interface of the blade and ice. This large pressure is responsible for melting the ice directly beneath the blade, allowing it to 'skate' smoothly over the surface on a thin film of water.

Pressure in Fluids

Atmospheric Pressure

Atmospheric, or air pressure, is the pressure produced by the combined weight of all the molecules in the atmosphere 'pressing down' on others beneath them. This means that at ground level, the whole weight of the atmosphere above acts on all beings and objects. Atmospheric pressure is equivalent to a force of about 100,000 N (100 kN) acting over an area of one square metre. Although this is a very large force, equivalent to a weight of 10 metric tons, its effects are not felt by objects or living things. This is because all things are in equilibrium, so that the pressure inside them is equal to, or balances, that outside. As a result of this, atmospheric pressure is ordinarily unnoticeable. The effects of atmospheric pressure can be felt or demonstrated though, in a number of ways.

When climbers ascend a mountain from ground level upwards, they experience a drop in pressure as they climb vertically. The higher they climb the more pronounced the pressure drop becomes. This is very apparent and causes symptoms like shortness of breath and light-headedness. At higher altitudes, there are less air molecules in the air; it is rarified, or lower in density, than at sea level.

This is because the higher one ascends, the less the amount, or weight of atmosphere above, pressing down and compressing air molecules below. Air is less dense; proportionately, there are less oxygen molecules present than at sea level.

The 'collapsing can' experiment demonstrates the effect of atmospheric pressure. A thin, sealed can made of tin or aluminium is de-pressurised by connecting its only outlet to a vacuum pump by a short length of tubing. The pump gradually evacuates the air from the inside of the can, until a near-vacuum is formed inside and the walls can no longer counteract the force of atmospheric pressure, now acting only on the outer surface of the can. The can is crushed completely under the force.

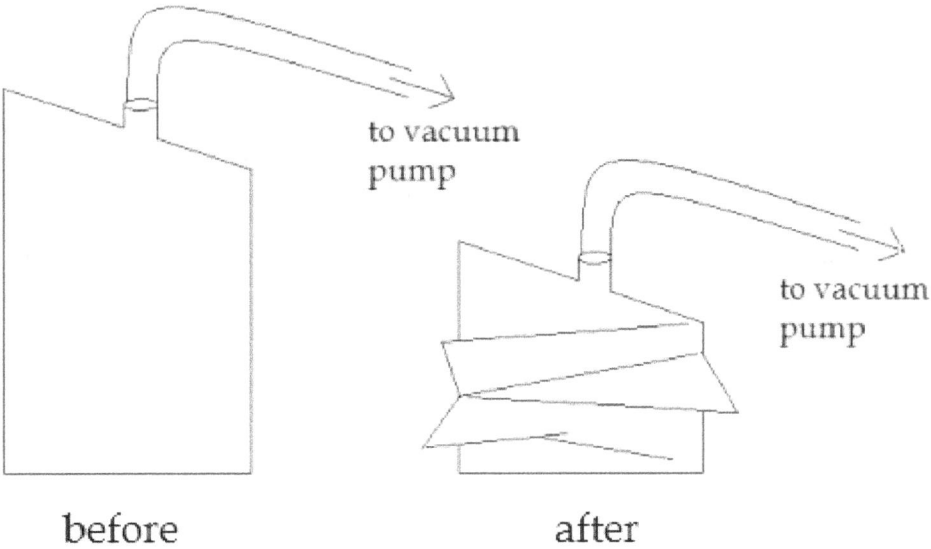

to vacuum pump

to vacuum pump

before   after

Pressure in Liquids

The height of a liquid column may be used to measure atmospheric pressure. In figure 7.2, a glass tube is completely filled with water, then inverted and placed in a beaker full of water. All of the water remains in the tube; its level does not drop at all.

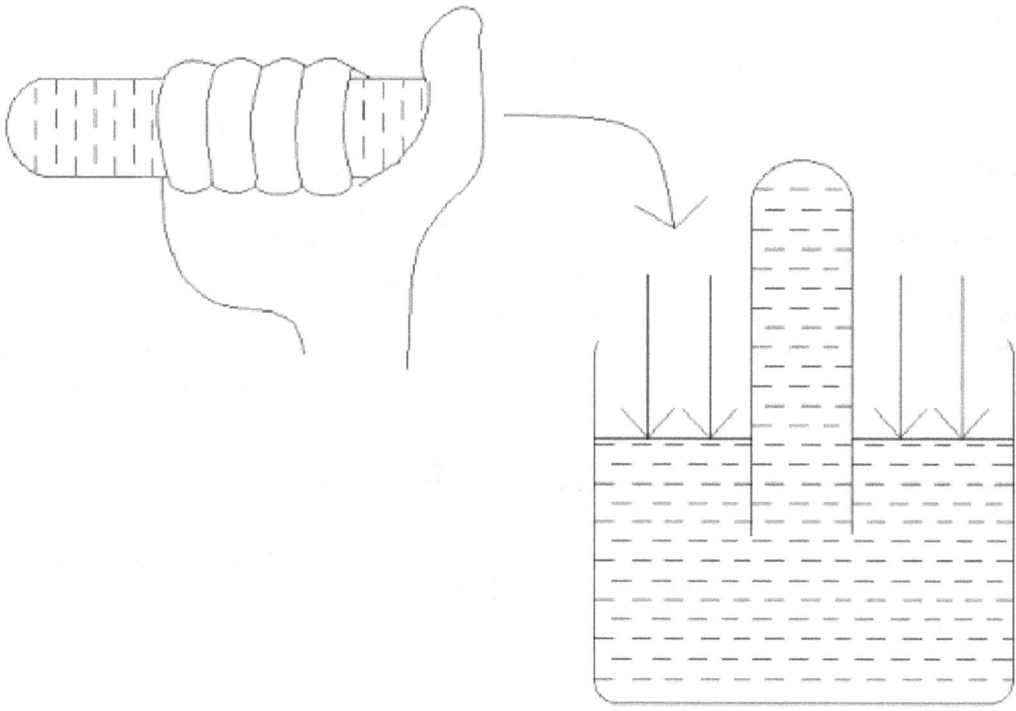

Figure 7.2

Illustrating the effect of atmospheric pressure

Atmospheric pressure is exerted on the free surface of the water. It exceeds the opposing pressure from the water column, which is acting to displace itself from the tube and increase the level of water in the beaker.

A mercury barometer uses this same principle in order to measure atmospheric pressure. If a 1m long glass tube is filled entirely with mercury, then inverted before being immersed in a beaker full of mercury, then the column's weight is enough to overcome atmospheric pressure. The level of the mercury column drops, forming a vacuum at the top of the tube. Notice that the **meniscus** formed at the glass-mercury interface is convex meaning that it bulges outwards from the liquid, in contrast to water, which forms a concave meniscus.

Mercury is a lot denser than water, it is 13.6 times heavier per unit volume. A column of mercury 760 mm high is required to balance atmospheric pressure under normal conditions - room temperature at sea-level. For the same effect with water, a tube nearly 10 m long would be needed; this is obviously impractical.

A Fortin barometer, which is commonly used in science to measure local atmospheric pressure for the purpose of experiments, operates in exactly this way. Instead of the mercury reservoir being open to the atmosphere though, it is sealed in a flexible pouch. A very accurate Vernier scale is needed to measure the exact pressure reading; as a small amount of mercury displacement represents a relatively large change in pressure.

The pressure in a liquid depends upon both its density and the specific depth beneath the surface. At a particular depth, the combined mass of all the liquid molecules above, plus atmospheric pressure, acts to create pressure. The pressure acts equally in all directions; this is a property of liquids which is employed as the basis of hydraulic systems. The denser the liquid, the greater the pressure at a given depth. Liquids such as alcohol or petrol are less dense than water, having a mass of about 700 kg/m^3 in comparison to 1000 kg/m^3 for water. This means that if two identical containers are filled with liquid, one with water and one with ethanol, then the pressure in the water container will be greater than that in the ethanol container at the same depth:

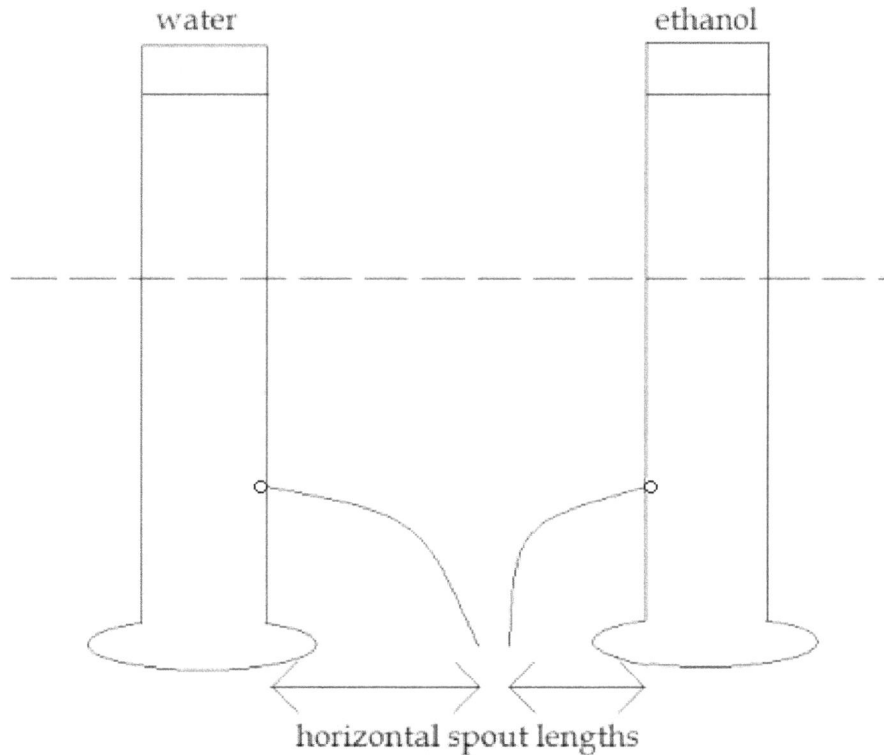

If a hole is drilled in the side of each container at an equal depth, then the spout of the liquid from the water container will be longer than that from the ethanol container. The weight of the water molecules above the hole is greater than the weight of ethanol molecules, even though their volumes are equal. That is, the water is denser, so it exerts more pressure. If more holes are now drilled in both containers, then the effect of depth on liquid pressure can be seen in each:

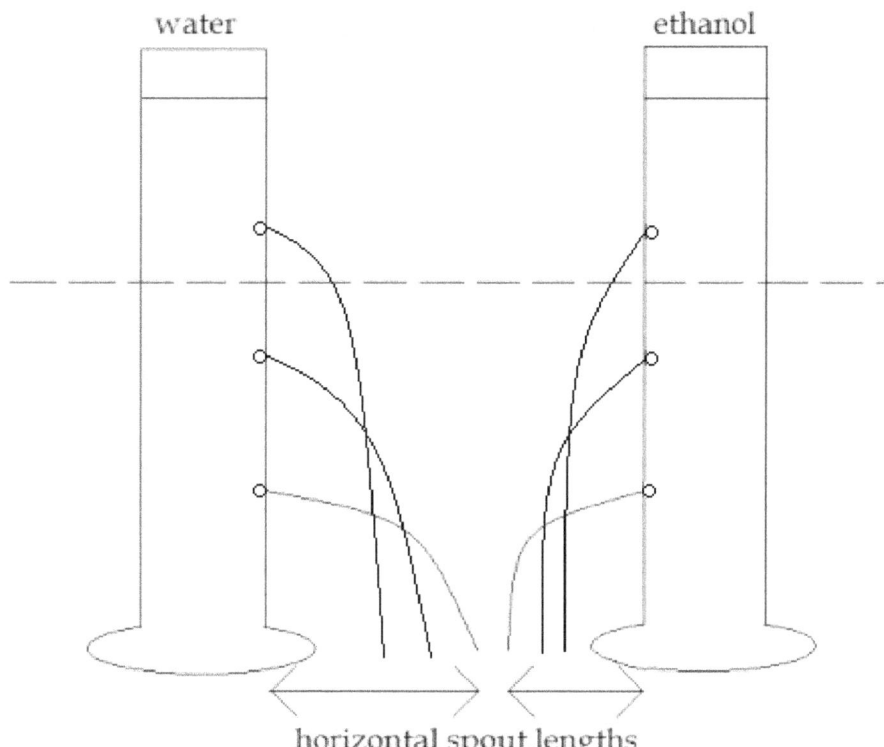

Note that the spout from the water container is always greater than that from the ethanol container when like depths are compared. But see that the spout length of both increases in the same way moving from the top to the bottom of the containers. This demonstrates the linear increase in pressure with depth true of all liquids.

In liquids, it is because of this linear increase in pressure with depth, coupled with the property that liquid pressure acts equally in all directions, that water damns are built thicker at the bottom than at the top:

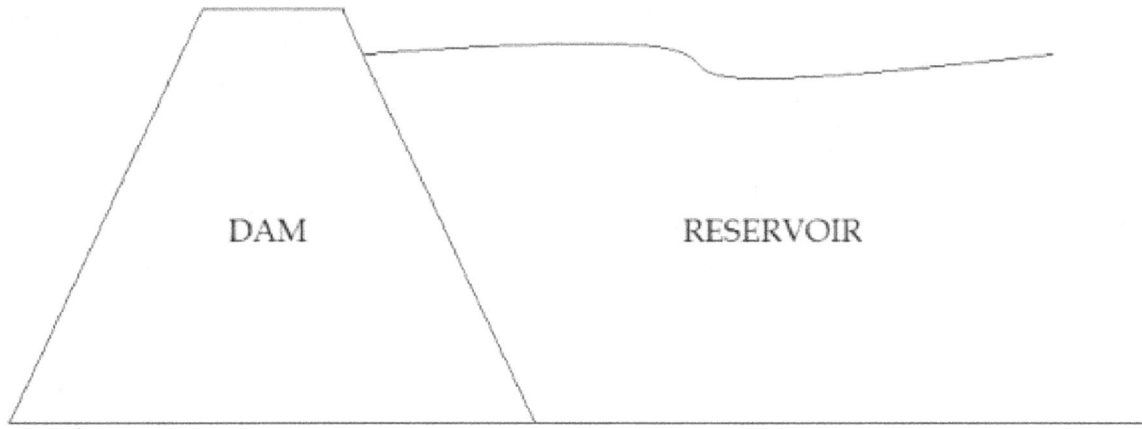

An expression for the pressure in a liquid at depth may be derived by examining the physical principles and forces involved:

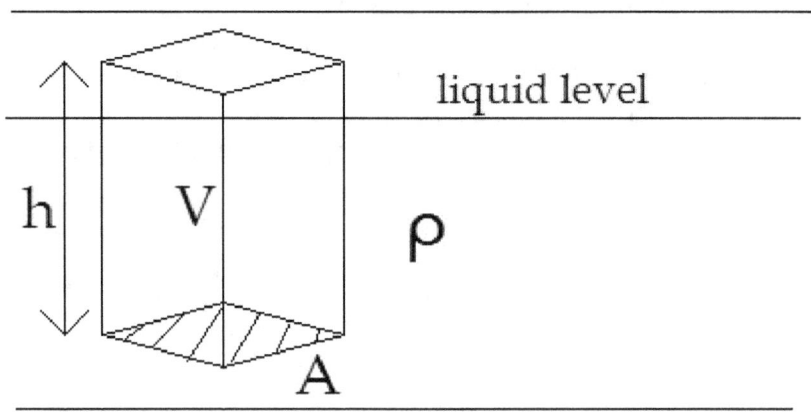

pressure p = force/area

p = weight of liquid in 'column'/cross-sectional area

p = (mass of liquid x gravitational pull)/cross-sectional area

p = (mass of liquid x g)/A

But mass of liquid = density x volume = ρ x V = ρ x (h x A)

Thus p = ρ h A g / A

i.e. p = ρ h g

That is, pressure = density x height of liquid 'column' x gravitational force

Hydraulics

Liquids transmit pressure equally in all directions, and any change in liquid pressure is trans-mitted instantly to all parts of the liquid.

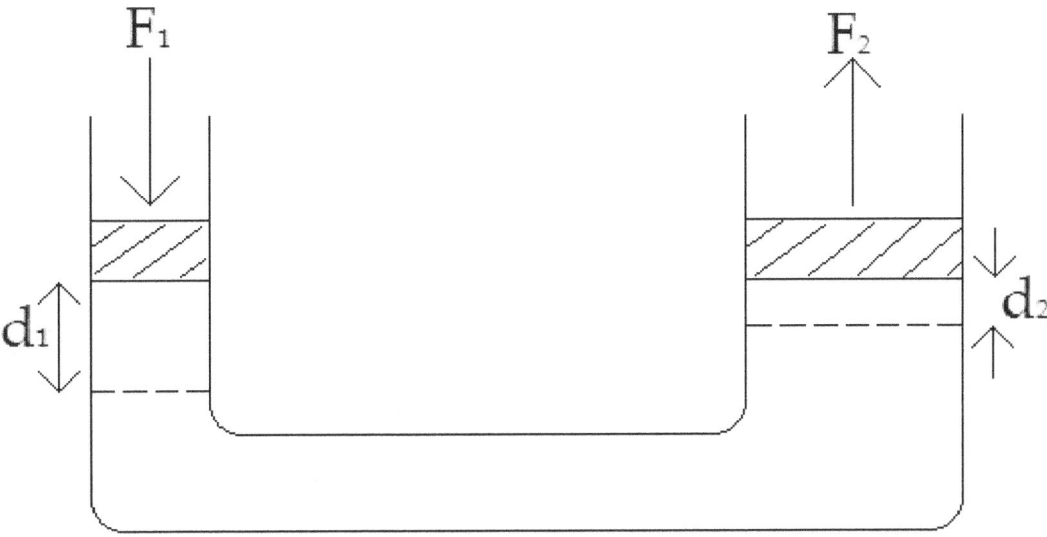

Figure 7.3

A small force F_1 pushes the piston a distance d_1, which moves the larger piston with a greater force F_2 but a smaller distance d_2 thanks to the instantaneous transfer of pressure

In the basic hydraulic system of figure 7.3, a force F_1 is applied to the smaller piston of area A_1, which produces a pressure beneath the surface of:

$$p = \frac{F_1}{A_1}$$

The pressure is transmitted equally and instantly through all parts of the liquid and is applied to the larger piston of Area A_2.

$$F_2 = p \times A_2 = \frac{F_1}{A_1} \times A_2$$

A_2 is greater than A_1, therefore a small force applied at A_1 is multiplied to become a large force applied at A_2. The force is greater, but energy is still conserved, since the distance moved by the piston at A_2 is correspondingly less also:

$$\text{Work done by } F_1 = \text{Work done by } F_2$$

$$F_1 \times d_1 = F_2 \times d_2$$

Examples

E.G. 1: A car's brakes operate on a hydraulic system.

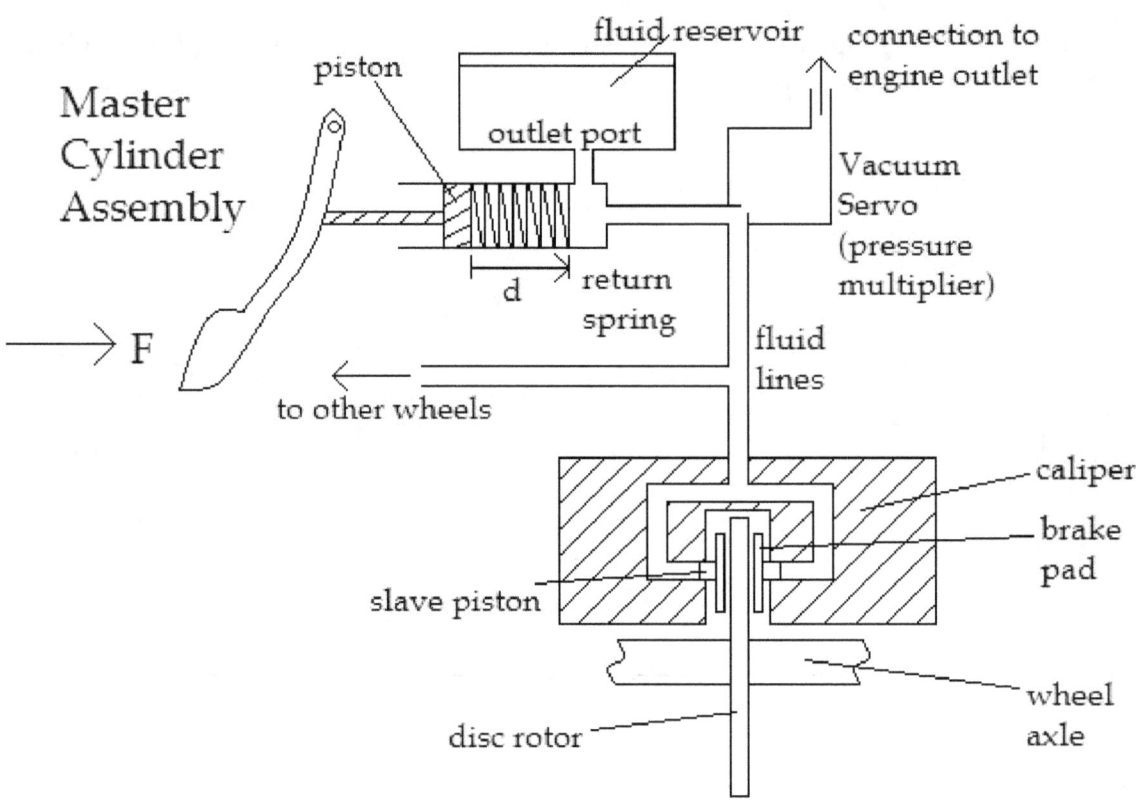

If F is 225 N, d = 105 mm and 15 cm^3 of fluid volume is displaced by the master cylinder piston, what is the pressure increase?

pressure = Force/ Area

force = 225 N.

$$\text{Area} = \frac{\text{Volume displacement}}{\text{Piston travel length}}$$

$$\text{Area} = \frac{15 \times (1/100)^3}{105 \times (1/1000)} = 1.42 \ldots \times 10^{-4} \text{ m}^2 = 1.42 \ldots \text{cm}^2$$

$$\text{Thus pressure} = \frac{225}{1.42 \times 10^{-4}} = 1.575 \text{ MPa}$$

The 'boost factor' provided by the vacuum servo is 5. If the area of contact of the disc pads with the rotor is 30 cm, find the pressure of the pads on the disc, given that the area of a caliper piston is 6 cm^2.

Force on pads = Pressure on piston x Piston area = (5 x 1.575 x 10^6) x (6 x (1/100)^2) = 4.725 kN

$$\text{Pressure on pads} = \frac{\text{Force on pads}}{\text{Area of pads}} = \frac{2 \times 4.725 \times 10^3}{30 \times (1/100)^2} = 3.15 \text{ MPa}$$

E.G. 2: A hydraulic press is used to compress bales of material into a flatter shape.

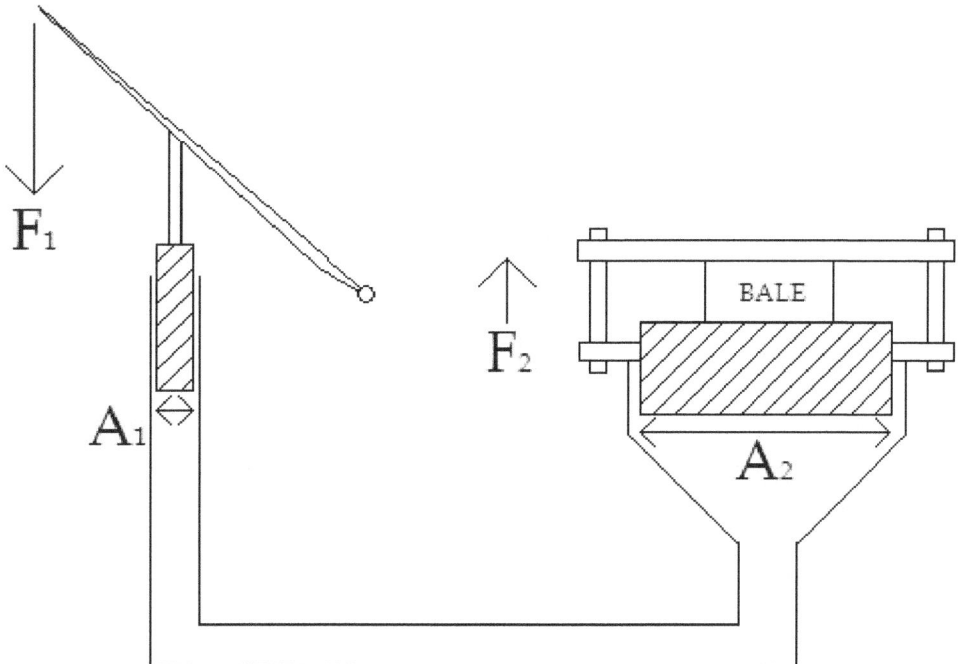

The bale is pressed tightly to the piston below by a pair of flange plates bolted together. An operator applies a force via a lever, pressurising the system by an amount F_1/A_1. This is transmitted through the press to the piston below, and the resultant force here is given by $F_2 = p \times A_2 = (F_1/A_1) \times A_2$. If the mechanical advantage of the press is 90%, so that the expected force F_2 is 0.9 that of the theoretical maximum, what force F_2 compresses the bale if a force F_1 of 200 N is applied at the operator's lever? (A_2 is 1500 cm^2 and A_1 is 150 cm^2.)

$$F_2 = 0.9 \times F_1 \times \frac{A_2}{A_1} = 0.9 \times 200 \times \frac{1500}{150} = 1.8 \text{ kN}$$

The initial area of the face of the cylindrical bale is 900 cm^2. If F_2 is 1.8 kN, what is the pressure generated at its interface with the piston?

$$\text{Pressure} = \frac{18\,00}{900 \times (1/100)^2} = 20 \text{ kPa}$$

Pressure in Gases

A manometer is an instrument used to measure pressure in gases. Typically, it is capable of measuring only small differences in pressure, either above or below atmospheric pressure. A water manometer can be used to measure the pressure in the gas mains, as an example of its operation. It consists of a U-shaped tube, open-ended on both sides, and three-quarters filled with water. The shorter end is connected to the gas outlet pipe. Without the gas flow turned on, the water in the manometer tube "finds its own level", meaning the top surfaces of the water in both arms of the tube are level with one another. Once the gas flows from the pipe to the manometer tube, it displaces the water column and the surface levels of the water move in opposite directions.

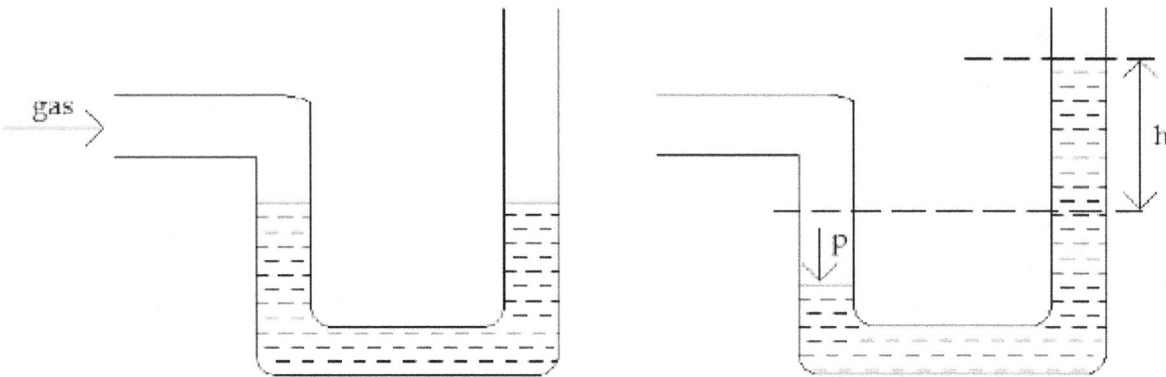

A manometer measuires only relative pressure, which is the pressure of the gas relative to its immediate surroundings, which are at atmospheric pressure. Relative pressure is normally referred to as gauge pressure. An absolute, or actual pressure reading is only obtained by adding this gauge reading to the atmospheric pressure. Strictly, the atmospheric pressure reading should be taken from a locally placed barometer for the purposes of experimentation, since it can vary considerably from a mean of around 760 mm Hg to lows of 740 mm and highs of 770 mm or more.

If the relative gas pressure is found to be equal to h = 11 cm of water, and the atmospheric pressure is taken to be 760 mm Hg, what is the absolute gas pressure in mm Hg?

$$P_{absolute} = P_{atmospheric} + P_{gauge}$$

P_{gauge} = 11 cm H_2O = 11 x(1/13.6) cm Hg (Mercury is 13.6 times more dense than water)

i.e. P_{gauge} = 0.808 ... cm Hg = 8 mm Hg (nearest mm)

Thus $P_{absolute}$ = 760 + 8 = 768 mm Hg

Changing the pressure of a fixed mass of gas at constant temperature causes a change in volume of the gas by an inversely proportional amount. If the volume of a fixed mass of gas is reduced isothermally (at constant temperature), then its gas pressure increases. A reduced volume means smaller distances between collisions and a greater frequency of collisions, causing the pressure increase.

For a fixed mass of gas at constant temperature, pressure is inversely proportional to volume:

$$P \, \alpha \, \frac{1}{V}$$

i.e. $P = \dfrac{k}{V}$ (k is a constant)

This is known as Boyle's Law:

$$PV = k \qquad\qquad (7.2)$$

Suppose a gas undergoes an isothermal change from an initial state where its properties (pressure, volume) are known. If, in its final state, either its pressure or volume is known, then the other can be found provided its initial state is fully specified.

Thus:

$$P_1V_1 = P_2V_2 \ \text{(for isothermal processes)}$$

E.G. A gas initially occupies a volume of 1.4 m^2 at $20\,^\circ$C. If its volume is now reduced to 0.4 m^2 in an isothermal process, find the change in pressure from an initial value of 2.1 MPa.

Isothermal process, so $P_1 V_1 = P_2 V_2$

$$(2.1 \text{ x } 10^6)(1.4) = P_2(0.4)$$

i.e.

$$P_2 = \frac{(2.1 \text{ x } 10^6)(1.4)}{0.4} = 7.35 \text{ MPa}$$

www.ingramcontent.com/pod-product-compliance
Lightning Source LLC
Chambersburg PA
CBHW081758170526
45167CB00008B/3241